CMS WORKSHOP LECTURES

VOLUME 3

THERMAL ANALYSIS IN CLAY SCIENCE

THE CLAY MINERALS SOCIETY

FREDERICK A. MUMPTON, EDITOR-IN-CHIEF

CMS WORKSHOP LECTURES

VOLUME 3

THERMAL ANALYSIS IN CLAY SCIENCE

by

D. L. Bish, Los Alamos National Laboratory, Los Alamos, New Mexico

C. J. Duffy, Los Alamos National Laboratory, Los Alamos, New Mexico

R. F. Giese, Jr., State University of New York at Buffalo, Buffalo, New York

Stephen Guggenheim, University of Illinois at Chicago, Chicago, Illinois

A. F. Koster van Groos, University of Illinois at Chicago, Chicago, Illinois

R. A. Ramik, Royal Ontario Museum, Toronto, Ontario, Canada

F. J. Wicks, Royal Ontario Museum, Toronto, Ontario, Canada

Technical Editors

J. W Stucki
Department of Agronomy
University of Illinois
Urbana, Illinois 61801

D. L. Bish
Los Alamos National Laboratory
ESS-1, Mail Stop D-469
P.O. Box 1663
Los Alamos, New Mexico 87545

Managing Editor

F. A. Mumpton
Department of the Earth Sciences
SUNY—College at Brockport
Brockport, New York 14420

Published by

The Clay Minerals Society
P.O. Box 4416
Boulder, CO 80306
USA

The Clay Minerals Society
P.O. Box 4416
Boulder, CO 80306
USA

For information on this and other volumes in the *CMS Workshop Lectures Series,* write The Clay Minerals Society, P.O. Box 12210, Boulder, Colorado 80303, U.S.A.

Citations of articles or chapters contained in this volume are properly referenced as follows:

Author (1990) Title: in *CMS Workshop Lectures, Vol. 3, Thermal Analysis in Clay Science,* J. W. Stucki, D. L. Bish, and F. A. Mumpton, eds., The Clay Minerals Society, Boulder, Colorado, 000-000.

TABLE OF CONTENTS

Preface vi

The Clay Minerals Society vii

Introduction
R. F. Giese 1

Precision Scanning Calorimetry of Clay Minerals and their Intercalates
R. F. Giese 9

High-Pressure Differential Thermal Analysis: Applications to Clay Minerals
A. F. Koster van Groos and Stephen Guggenheim 49

Thermogravimetric Analysis of Minerals
D. L. Bish and C. J. Duffy 95

Vacuum Thermogravimetric Analysis and Evolved Gas Analysis by Mass Spectrometry
F. J. Wicks and R. A. Ramik 190

Mineral Index
J. W. Stucki 191

PREFACE

Since 1982, The Clay Minerals Society through its Continuing Education Committee has sponsored workshops on various topics of interest to clay scientists and technologists in conjunction with its annual meetings. In 1982, in Hilo, Hawaii, the workshop subject was sample preparation techniques; in 1983, in Buffalo, New York, infrared spectroscopic techniques were treated; and in 1984, in Baton Rouge, Louisiana, the use of clay minerals in petroleum well logging was explored. In 1985, in Denver, Colorado, the subject was the quantitative mineral analysis of clays, and in 1986, in Jackson, Mississippi, the electron-optical methods in clay science were examined. Lectures from these last two workshops have been published as *CMS Workshop Lectures, Volume 1, Quantitative Mineral Analysis of Clays* and *CMS Workshop Lectures, Volume 2, Electron-Optical Methods in Clay Science* and are available from The Clay Minerals Society.

In 1987, at its annual meeting in Socorro, New Mexico, a workshop on Advanced Methods of Thermal Analysis of Clay Minerals was held on October 17 and was attended by 70 participants. The workshop was organized and conducted by R. F. Giese, Jr., Department of Geological Sciences, State University of New York at Buffalo, convener. Major contributors to the workshop included R. H. April, S. H. Blassingame, D. G. Horton, R. C. Jones, and D. R. Pevear from the Continuing Education Committee; D. L. Bish, C. J. Duffy, W. N. Frazier, R. F. Giese, Jr., Stephen Guggenheim, A. F. Koster van Groos, and F. L. Wicks, lecturers; and Du Pont de Nemours Corporation, which provided instrumentation.

The present volume is the third in the Society's *CMS Workshop Lectures* series and is based on updated versions of several of the lectures presented by the 1987 workshop. The subjects treated herein provide a sound basis for the use of modern thermal analytical techniques in the study of clays. This volume, as was its precursor Workshop, is intended for students who wish to begin their thermal investigations of clay minerals and for professionals who want to renew their acquaintance with this important subject. Each contribution has benefited from peer review and represents in its own right a valuable contribution to the scientific literature. The efforts of numerous technical referees are hereby acknowledged, as are the managing editor contributions of F. A. Mumpton and the clerical assistance of A. G. Mumpton.

J. W. Stucki D. L. Bish

THE CLAY MINERALS SOCIETY

The Clay Minerals Society was organized in 1963 to stimulate research and to disseminate information relating to all aspects of the science and technology of clays and other fine-grained minerals. It sponsors annual Clay Minerals Conferences where research and invited papers are presented in technical sessions and special symposia and organizes field trips to important occurrences of clays in all parts of North America and to industrial sites of clay production and application. In conjunction with its annual meetings, workshops are held on technical subjects of interest to clay researchers and technologists.

The Clay Minerals Society publishes bimonthly *Clays and Clay Minerals*, which is one of the leading international Journals in the field of clay science. *Clays and Clay Minerals* presents the latest scientific investigations in all areas of the field and from all parts of the world, along with timely review articles and announcements of new publications on clays and other fine-grained minerals. Its half-tone illustrations of electron micrographs and other photographs are among the best scientific reproductions to be found.

The Society also sponsors a Source Clays project designed to provide homogeneous clay samples for research and teaching purposes and offers clay-related publications to its members at a discount over publishers' list prices.

Its multidiscipline membership includes agronomists, surface chemists, physicists, geologists, mineralogists, geochemists, material scientists, soil scientists, crystallographers, sedimentologists, economic geologists, colloid chemists, ceramists, rheologists, petroleum engineers, and geotechnical engineers, thereby offering members the opportunity to exchange ideas and results with fellow researchers having widely different backgrounds and expertise.

For information on joining the Society and subscribing to *Clays and Clay Minerals*, write to:

The Clay Minerals Society
P.O. Box 4416
Boulder, CO 80306
USA

INTRODUCTION

R. F. GIESE, JR.

Department of Geological Sciences
State University of New York at Buffalo
Amherst, New York 14226

Thermal analysis involves the observation of a physical property of a sample and how that property changes in response to a change in temperature. Thus, the essence of this group of techniques includes the measurement of a physical property, e.g., mass, temperature, and volume, and the control of temperature. Inasmuch as heating objects is a very ancient practice, one should not be surprised that the first observations of the *response* of certain materials to heat were made quite some time ago. Such observations might be considered as a form of thermal analysis (Mackenzie, 1981), but serious investigations required that the temperature be known with reasonable accuracy. Temperature measurements, especially of a solid material that is being heated rapidly, was first accomplished with a thermocouple. Two events, then, mark the beginning of thermal analysis. The first was the invention of the thermocouple. This led directly to the study of the thermal properties of a group of clay minerals. In fact, thermal analysis, in the modern sense, started with a simple description: "*Si l'on échauffe rapidement une petite quantité d'argile, il se produit, au moment de la déshydratation, un relentissement dans l'élevation de température...*" (if one heats rapidly a small quantity of clay, there is, at the point of dehydration, a slowing in the increase in temperature...) (Le Chatelier, 1887). The temperature at which the dehydration occurred was determined for each of the clay minerals examined by Le Chatelier, and he pointed out that the temperature at which dehydration occurred could be used to distinguish between and to identify different clay minerals.

Le Chatelier's short paper encompassed the major elements of thermal analysis: heating of a sample in a controlled manner, coupled with the measurement of a physical property of the sample (here, the change of temperature of the sample due to loss of water). The final development that was needed for the accurate measurement of thermal events came with the joining of two thermocouples in a reversed sense so that the total voltage produced by the pair reflected the difference in temperature of the two (Roberts-Austen, 1899). This invention was the basis for differential thermal analysis (DTA), in which one thermocouple is placed in contact with the sample and the other thermocouple is placed in a thermally inert reference material. DTA quickly became the primary tool for investigating the thermal properties and reactions of materials for about the next 60 years. DTA experiments are more complicated than were those of Le Chatelier; but the advantage of the DTA system is that the voltage produced by the differential arrangement of the two thermocouples is directly related to the thermal reaction of the sample.

The initial report of Le Chatelier was followed by studies of the thermal reactions of a wide variety of clay minerals (e.g., Wallach, 1923; Satoh, 1918; Orcel and Caillère, 1933; Granger, 1934; Insley and Ewell, 1935) in an attempt to identify the thermal reactions of individual species. Such studies led naturally to attempts to use the thermal reactions of clay minerals first to identify unknown samples and then to determine quantitatively the composition of mixtures (Orcel, 1935; Norton, 1939). The tantalizing prospect of using DTA for the quantitative analysis of clay-mineral mixtures inspired a number of workers, especially in the United States, to apply this technique to both artificial mixtures and real samples. The attempt to perfect thermal analysis as a quantitative tool became the major focus of research in the 1930s and 1940s (e.g., Norton, 1939; Grim and Rowland, 1942; Grim, 1947; Kerr and Kulp, 1948). In addition, DTA was applied to a host of non-clay minerals, including, eventually, very reactive materials, such as sulfides and borates.

During this period, a typical DTA apparatus, often not of commercial origin, consisted of a large metal block containing one or more holes drilled into it for the sample and the reference materials (Kerr and Kulp, 1948). The thermocouples were introduced by means of a smaller hole in the base of the block. The entire assembly was then placed in a furnace. With this arrangement, a number of very difficult problems soon became apparent that had to be solved before thermal analysis could be routinely and widely used for quantitative mineral analysis. These problems were of three types: (1) understanding the heat flow in the sample and reference chambers during heating and, especially, when the sample was undergoing a thermal reaction; (2) controlling the experimental parameters (e.g., sample size, sample dilution, particle size, heating rate, escaping vapor); and (3) sorting out overlapping peaks of thermal events on the chart paper used to record the differential voltage.

Modeling the heat flow in the sample during linear heating of a metal block, such as described above, can be accomplished (Eriksson, 1953) if the sample does not undergo a thermal reaction (the so-called base-line model). The basis for quantitative analysis is that the area under the peak of a thermal event is proportional to the mass of the reacting sample and the enthalpy of the reaction. Several workers have attempted to demonstrate this (see Sewell and Honeyborne, 1957, for a discussion of these works). Such studies showed that the peak area is independent of the heating rate, the heat capacity of the sample, and the rate of reaction; but that the area depends on a number of other factors, such as the thermal conductivities of the sample and the reference, the furnace design, and the flow of heat from the sample block (Sewell and Honeyborne, 1957). In addition, and this problem is very difficult to incorporate into the mathematical models of heat flow, one must account for the role played by the thermocouple wires that lead directly from the sample and reference cavities to the exterior of the furnace.

One might expect that sample preparation was not a major problem; but in practice, this was not the case. Duplication of the conditions of an experiment from run to run was, in fact, difficult. The major problems were: (1) the inability to place the thermocouples in the sample and reference material in exactly the same position for each experiment, and (2) the difficulty of packing the powdered sample (usually diluted with an inert material) and the reference around the thermocouple in a reproducible manner. Variations in either the position of the thermocouple or in the loading of the sample and reference strongly influenced the resulting thermograms.

Finally, in mixtures of minerals the exothermic and endothermic reactions often overlapped. The major impediment to resolving this problem was that, although thermal peaks should have a simple and predictable shape so that overlapping peaks could be deconvoluted, as is carried out in X-ray powder diffraction experiments, in practice, thermal peaks commonly have shapes that deviate markedly from the ideal. In such situations, quantitative analysis is impossible. With the realization that quantitative applications of DTA were not feasible, except in very special situations, the use of thermal analysis for mineral studies declined. In retrospect, the application of thermal analysis to quantitative ends seemingly was inappropriate and, unfortunately, damaged the image of the technique for many mineralogists.

In 1964, a new approach to thermal analysis appeared with the development of a technique that was not based on comparing the temperature difference between a sample and a reference, but, rather, was based on measuring the electrical power needed to keep the sample and the reference at precisely the sample temperature (Watson *et al.*, 1964). The fact that the reference and the sample were maintained at the same temperature regardless of the heating conditions of the experiment (i.e., linear heating or cooling or isothermal) simplified the heat-flow conditions. In addition, the sample and reference were placed in small separate cups of low thermal mass, each having a separate heating element and an electrical resistance thermometer. Thus, the temperature of both was easily and accurately controllable. This apparatus was termed a differential scanning calorimeter (DSC) and was designed for low to moderate temperatures ($<900°C$). For higher temperatures, DTA remains the only readily available technique, although DSC is currently being extended to more demanding regimes (e.g., Sandrock, 1982).

One of the more significant modifications to the DTA and DSC methods was the introduction of low- and high-pressure capability. DTA instruments were apparently modified for high-pressure operation as early as 1923 (see references in Sawade *et al.*, 1987); however, this technique has only recently become popular. Many manufacturers of thermal analysis equipment now have high-pressure DTA or DSC instruments available (e.g., Levy *et al.*, 1970). High-pressure instruments have been described that use gases, liquids, or solids as pressure media, with attainable pressures from well below one atmosphere to in excess of 25,000 atmospheres. The instruments have been used to study a variety of inorganic and organic reactions and phase transitions. Numerous decarbonation, dehydration/hydration, melting, and oxidation/reduction reactions have been studied at high pressures, and Sawada *et al.* (1987) documented much of the literature on this subject. The ability to study reactions *in situ* over a wide range of temperature and pressure is clearly a valuable tool in the study of clays and clay minerals.

The latest major improvement to DTA and DSC instrumentation occurred in the early 1980s with the incorporation of computers into the control of the experiments and the collection of data (Brennan and DiVito, 1982). In addition, the application of thermal analysis to a wider range of materials, especially plastics, ceramics, and composites, increased the market for thermal analysis equipment, resulting in improved instrumentation.

Thermal gravimetric analysis (TGA) is nearly as old as differential thermal analysis. The first study of the change in mass of a sample as a function of temperature was published by Nernst and Risenfeld (1903). They used a simple quartz torsion balance to examine the effects of heating on various minerals. Since then, numerous commercial instruments have appeared, based on microbalances modified to allow heating of the sample. Wendlandt (1986) summarized the history of the development of TGA and included numerous references on the subject. As with DTA, recent developments have contributed greatly to the flexibility and utility of TGA. First, the ability to control the temperature and its rate of change accurately made TGA a precise technique. No longer is it necessary to control temperature manually or to use poorly known heating rates. Second, the addition of microcomputer control of the instrumentation has greatly expanded the versatility and power of these methods. A computer-controlled instrument provides great flexibility and reproducibility in the execution of experiments, allowing multiple heating rates to be used over pre-determined temperature ranges, isothermal data collection, and repetitive experiments, which produce data as the temperature is increased and then decreased. Moreover, data analysis with a computer-automated instrument is rapid, often automatic, and much more versatile than with manual, chart-recorded instruments. With the use of digital thermal analytical data, the possibilities of complex data analysis are virtually limitless.

In addition to these fundamental changes in instrumentation, several ancillary techniques have been attached to thermal analytical equipment and have expanded the number of potential applications and types of information available from a sample. Numerous studies have employed a combination of a thermogravimetric analyzer and a mass spectrometer to obtain time-resolved information on both the quantity and the molecular weight of gas species evolved from a sample. Gibson and Johnson (1972) and Gibson (1973) described some of the first applications of mass spectrometry coupled to thermogravimetric analysis of geologic samples, and Langer (1986) described the early developments in the analysis of evolved gases by a mass spectrometer connected to thermal analytical devices. Today, several manufacturers of mass spectrometers offer equipment that can easily be coupled to TGA instruments.

A more recent innovation in this area is the use of a Fourier-transform infrared spectrometer (FT-IR) to analyze the gases produced in a thermogravimetric analyzer. Unlike a mass spectrometer, which provided information only on the molecular weight of evolved gaseous species, an FT-IR apparatus can provide precise and detailed information on the species being evolved. Coupling either a mass spectrometer or an FT-IR instrument to conventional TGA equipment can provide valuable information beyond that normally obtained, information that can facilitate a more complete understanding of thermal stability and decomposition reactions.

These and other developments have produced analytical equipment that is more sensitive and versatile than previously available. Unfortunately, while these improvements in thermal analysis were being made, the mineralogical community has lost interest in the technique. One of the purposes of this volume and the workshop on which it is based is to present some of the more recent applications of thermal analysis to clay scientists in the hope that they will take advantage of what is a rebirth of this valuable analytical technique.

REFERENCES CITED

Brennen, W. P. and DiVito, M. P. (1982) Computers in the plastics laboratory: Trends and prospects: Paper 484 presented at 1982 Pittsburgh Conf. Anal. Chem. Appl. Spectr., Atlantic City, New Jersey.

Eriksson, E. (1953) Problems of heat flow in differential thermal analysis: I.: *LantbrHögsk. Ann.* **19,** 127-143.

Gibson, E. K., Jr. (1973) Thermal analysis-mass spectrometer computer system and its application to the evolved gas analysis of Green River shale and lunar soil samples: *Thermochim. Acta* **5,** 243-255.

Gibson, E. K., Jr. and Johnson, S. M. (1972) Thermogravimetric-quadrupole mass spectrometric analysis of geochemical samples: *Thermochim. Acta* **4,** 49-56.

Granger, A. (1934) Thermal analysis of clay: *La Ceramique* **37,** 58-64.

Grim, R. E. (1947) Differential thermal curves of prepared mixtures of clay minerals: *Amer. Mineral.* **32,** 493-496.

Grim, R. E. and Rowland, R. A. (1942) Differential thermal analysis of clay minerals and other hydrous materials: *Amer. Mineral.* **27,** 746-761, 801-818.

Insley, H. and Ewell, R. H. (1935) Thermal behavior of the kaolin minerals: *J. Res. Natl. Bur. Stnd.* **15,** 615-627.

Kerr, P. F. and Kulp, J. L. (1948) Multiple differential thermal analysis: *Amer. Mineral.* **33,** 387-419.

Langer, H. G. (1986) Early developments of EGA by mass spectrometry: *Thermochim. Acta* **100,** 187-202.

Le Chatelier, M. H. (1887) De l'action de la chaleur sur les argiles: *Bull. Soc. Min.* **10,** 204-211.

Levy, P. F., Nieuweboer, G., and Semanski, L. C. (1970) Pressure differential scanning calorimetry: *Thermochim. Acta* **1,** 429-439.

Mackenzie, R. C. (1981) Thermoanalytical methods in clay studies: in *Advanced Techniques for Clay Mineral Analysis,* J. J. Fripiat, ed., Elsevier, Amsterdam, 5-29.

Nernst, W. and Riesenfeld, E. H. (1903) Über quantitative Gewichtsanalyse mit sehr kleinen Substanzmengen: *Chem. Berichte* **36,** 2086-2093.

Norton, F. H. (1939) Critical study of the differential thermal method for the identification of clay minerals: *J. Amer. Cer. Soc.* **22,** 54-63.

Orcel, J. (1935) Differential thermal analysis in the determination of the constituents of clays, laterites, and bauxites: *Cong. Int. Mines, Met. Geol., Appliq., VII Sess., Paris,* 359-373.

Orcel, J. and Caillère, S. (1933) L'analyse thermique différentielle des argiles á montmorillonites (bentonites): *C.R. Acad. Sci., Paris* **197,** 774-777.

Roberts-Austen, W. C. (1989) Fifth report to alloys research committee: in *Proc. Insts. Mech. Engrs.* 35-102.

Sandrock, R. (1982) High-pressure high-temperature differential scanning calorimetry: *Rev. Sci. Instrum.* **53,** 1079-1081.

Satoh, S. (1918) On the endo- and exothermal change of the kaolinite in Japan: *Kogyo-Kwagaku-Zasshi* **21,** 631-648.

Sawada, Y., Henmi, H., Mizutani, N., and Kato, Masanori (1987) Differential thermal analysis under high-pressure gas atmospheres: Applications to materials science and engineering. A review: *Thermochim. Acta* **121,** 21-37.

Sewell, E. C. and Honeyborne, D. B. (1957) Theory and quantitative use: in *Differential Thermal Investigation of Clays,* R. C. Mackenzie, ed., Mineralogical Society, London, 65-97.

Wallach, H. (1913) Analyse thermique des argiles: *Compt. Rend.* **157,** 48-50.

Watson, E. S., O'Neill, M. J., Justin, J., and Brenner, N. (1964) Differential scanning calorimetry for quantitative differential thermal analysis: *Anal. Chem.* **36,** 1233-1238.

Wendlandt, W. W. (1986) *Thermal Analysis:* Wiley, New York, 814 pp.

DIFFERENTIAL SCANNING CALORIMETRY OF CLAY MINERALS AND THEIR INTERCALATES

R. F. GIESE, Jr.

CONTENTS

Introduction	10
Theory of Decomposition of Solids	11
Decomposition events	11
Clay mineral intercalates	13
Experimental Methods	14
Thermogravimetric analysis	14
Differential scanning calorimetry	15
Measurements	17
Sample preparation and operation	18
Sub-ambient operation	19
Kinetics of Deintercalation	19
General	19
Inhomogeneous rate laws	20
Contracting circle	21
Avrami-Erofeev law	22
Differential scanning calorimetry	22
Activation energy	24
Determining the rate law	25
Examples: Deintercalation	27
Experimental conditions	28
Computations	29
Heat Capacity Measurements	31
Examples: Heat Capacity	34
Single material	35
Experimental conditions	35
Computations	36
Intercalated material	37
Experimental conditions	38
Computations	39
Discussion	40
Deintercalation of N-methyl formamide	41
Samples	41
Deintercalation results	42
Heat capacity of water intercalated in kaolinite	42
Summary	46
References Cited	46

DIFFERENTIAL SCANNING CALORIMETRY OF CLAY MINERALS AND THEIR INTERCALATES

R. F. GIESE, JR.

Department of Geological Sciences
State University of New York at Buffalo
Amherst, New York 14226

INTRODUCTION

The field of thermal analysis covers a wide variety of techniques, each of which looks at specific characteristics of a sample as the temperature is raised or lowered at a controlled rate. This chapter describes many of the events that take place when a clay mineral that has been intercalated by a molecular species, e.g., water or various organic molecules, is heated or cooled. If a clay intercalate is heated, it decomposes (at a given rate) into two phases: the deintercalated clay and a vapor of the intercalated species. The rate of decomposition, the activation energy, and the rate law are all of interest.

Cooling the sample below ambient temperatures allows the heat capacity, C_p, of the intercalated clay to be examined and, by suitable manipulations described below, the C_p of the intercalated phase to be determined, at least approximately. These two heat capacities are of interest in the examination of the interactions of the guest molecules with the adjacent clay mineral surfaces (Lipsicas *et al.*, 1986).

The chief purpose of the following discussion is to describe several thermal analytical techniques that are applicable to clay intercalates, the theory behind these techniques, the mathematical manipulations and calculations that follow the measurements, and, to a limited extent, how the results might be interpreted. This last point will not be stressed, because the present discussion is about methods; the citations in the text, however, allow the reader to consult references in which possible interpretations are outlined extensively.

The experimental techniques discussed in this chapter refer specifically to the Perkin-Elmer DSC-2 differential scanning calorimeter, the instrument used by the author, but several other instruments of this type are also commercially available. It is not feasible to take into account all the idiosyncrasies of each of these other machines when describing how a particular analysis is done. With only one exception (described below), however, all differential scanning calorimeters work essentially the same way, and the descriptions in this chapter should be generally applicable to all such instruments.

THEORY OF DECOMPOSITION OF SOLIDS

The decomposition of solids has been studied from a number of aspects, but principally with respect to the dehydration of hydrated solids and the decomposition of unstable materials (Mackenzie, 1957; Young, 1966; Wendlandt, 1986). In a narrow sense, however, the thermal decomposition of a clay intercalate is different from either of these processes in that: (1) with intercalates, the process is more akin to the desorption of a weakly adsorbed volatile phase, (2) a reaction involving the breaking of strong chemical bonds is not normally involved, and (3) the intercalated species play no central role in the clay structure comparable to that of the water of hydration in a mineral such as gypsum. Because the mathematical description of the rate at which an event occurs is independent of the nature of the processes involved, the kinetics of the deintercalation of clay minerals can be described by the methodology developed for the decomposition of solids.

Decomposition events

The fundamental quantity of interest in a decomposition reaction is the variation of the fraction of sample, commonly termed "α", which has decomposed as a function of either time or temperature. Plots of "α" vs. time generally yield a sigmoidal curve, which has certain features characteristic of the particular experiment (Young, 1966). Such a curve is shown in Figure 1 in which the different segments are numbered. A curve of this kind, for example, is traced during a thermogravimetric run in which the temperature is increased linearly or in which the temperature is held constant at some value above the point at which the solid becomes unstable. The following discussion is largely taken from Young (1966), and his nomenclature has been used.

Segment 1 in Figure 1 is the initial, rapid evolution of vapor from the sample. The amounts typically observed are small, generally less than a few percent. For intercalated clays, observation of this feature, particularly if the amount exceeds a few percent, means that the clay has some surface-adsorbed material in excess of that needed for complete intercalation (Adams, 1978). Proper treatment of the sample will remove this excess. The initial event is followed by the "induction period", characterized by a slow evolution of vapor (segment 2 in Figure 1). The induction period is followed by the "acceleratory period" (segment 3 in Figure 1), which is the period in which the rate of the decomposition increases, reaching a maximum at the transition to the next phase, the "decay period" (segment 4 in Figure 1). The decay period marks the cessation of the decomposition, assuming that all material capable of decomposition has in fact decomposed.

Clearly, each of these stages represents different events taking place in the sample, and each probably has its own reaction rate and even its own rate law. An attempt could be made to determine the parameters of each, but in practice, most interest is centered on the acceleratory and decay stages. As shown below, if all the

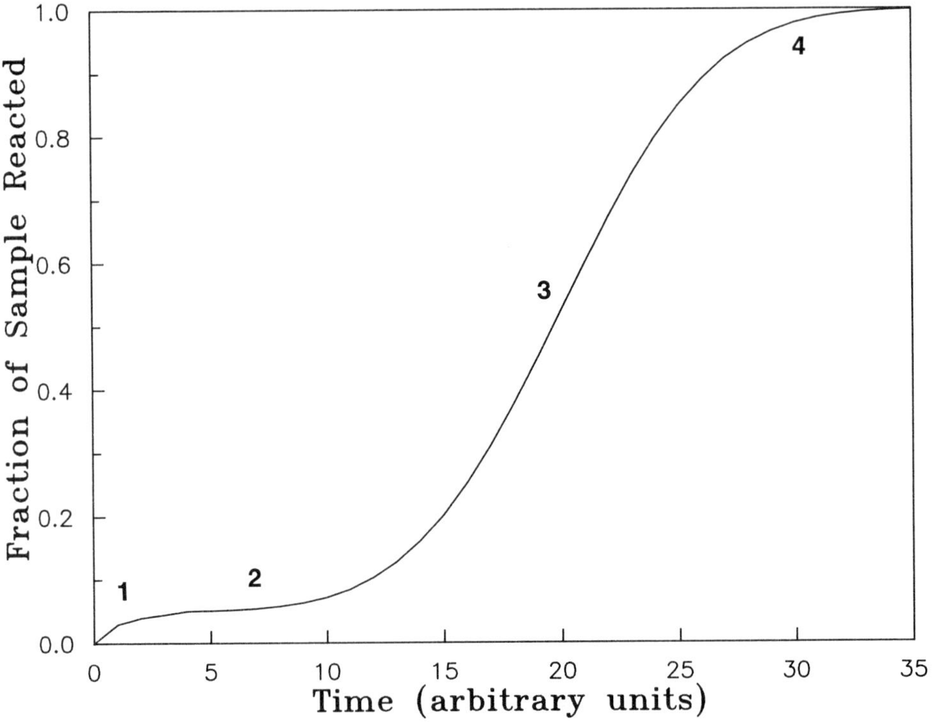

Figure 1. An idealized representation of weight loss (α = the extent of reaction) of a material that is decomposing at constant temperature. Region 1 is a transient phase often missing in real experiments; region 2 is the "induction period"; region 3 is the "acceleratory period"; region 4 is the "decay period".

data from an experiment are analyzed using the Arrhenius equation, i.e, "α" from 0 to 1, the rate constants, k_t, at the beginning and end of the experiment will not fall on the straight line defined by the rate law. These extremes typically are ignored, because they represent only a small fraction of the decomposition and also because they represent data having a large experimental error.

In mechanistic terms, the decomposition of a solid involves the nucleation of the decomposition product, the growth of the nuclei, and the evolution of the mobile phase or phases (Young, 1966). The formation of nuclei has been studied using single crystals of hydrated salts (e.g., Young, 1966); but nucleation in hydrated salts is unlikely to proceed in the same manner as nucleation during the decomposition of intercalated clays. For the latter, the nuclei must represent some sort of change from a structure in which an organic molecule, for example, is more or less rigidly held between the faces of opposing clay layers, to a second structure having a mobile molecule, perhaps associated with other similar molecules, which are actively leaving the clay matrix.

Possibly, the loss of the guest phase after nucleation is controlled by diffusion and, thus, this event should be seen clearly in the kinetics law describing the accelera-tory and decay stages; this, however, has yet to be observed (see Adams and Waltl, 1980, for a discussion). Thus, the important events for clay intercalates seem to be nucleation and growth of the nuclei.

Clay mineral intercalates

Intercalation is defined as the introduction, by suitable treatment, of a foreign molecular or ionic material between the layers of a clay mineral. The mechanisms that drive this introduction are varied, but include: ion exchange, coordination of a pre-existing interlayer cation by guest molecules, and hydrogen bonding between a guest molecule and a clay surface. Intercalation has been recognized for a long time, and the types of materials that will intercalate into different clays are discussed in a number of standard reviews (Theng, 1974; MacEwan and Wilson, 1980).

The mechanism by which a molecule enters between the layers of a clay are complex and are only partly understood. Clearly, the introduction of an external species must, locally at least, distort the clay layers that are pushed apart to make room for the molecules. This process involves energy. The change of environment of the molecule as it leaves the bulk to occupy a volume that is delimited by the surfaces of the clay layers involves a change in entropy. Finally, interactions should take place between the guest molecule and the interlayer cation (if there are any), and/or the surface oxygens of the clay layers, and/or other neighboring mole-cules—all of which have associated energy terms. The end result, ideally at least, is a layered structure containing planes of guest molecules alternating with the clay layers. Because intercalation involves an energy barrier (or barriers), the rate of intercalation should be influenced by temperature, and temperature dependence should translate into an activation energy. In a few instances, such an experiment has been carried out (Hach-Ali and Weiss, 1969).

The formation of an intercalated clay involves interactions between the guest species and the host clay. The reverse process, the decomposition of the intercalate, is of fundamental interest also because it is controlled by the stability of the interca-late and by the strength and nature of the interactions between the guest species and the clay surfaces. If the guest molecule has an appreciable vapor pressure, heating will decompose the intercalated clay into a clay mineral plus an organic phase or water vapor, which is lost. These processes of intercalation and deintercalation, while reversible, probably do not employ the same mechanisms.

EXPERIMENTAL METHODS

Two general ways of performing experiments are employed to determine kinetics of a reaction or decomposition: *isothermal* and *dynamic*. As the name indicates, isothermal refers to an experimental procedure in which the sample is held at a constant temperature and in which the reaction is monitored as a function of time, whereas dynamic covers experiments in which the sample is heated at a pre-determined rate. Both have advantages and difficulties, some of which will be enumerated in the following sections. Except where noted, the following discussions refer specifically to the dynamic mode, because this has been the more common type of experiment.

Of all the thermal analytical techniques available today (see Mackenzie, 1981, for a review of clay-related methods), only two will be discussed in this contribution: thermogravimetric analysis (TGA) and differential scanning calorimetry (DSC), and of these, DSC will receive more attention. The following is a brief survey of the two techniques and is meant to serve as background material for use in later sections.

Thermogravimetric analysis

Several phenomena initiated by heating at a controlled rate or at a fixed temperature above ambient involve a change (usually a loss) in weight (e.g., dehydration, desorption, decomposition, oxidation). TGA is essentially a means of observing the weight of a sample as a function of temperature (dynamic heating) or time (isothermal heating). In its simplest form, the equipment consists of a sensitive balance and a furnace arranged such that the sample holder sits inside the furnace. A thermocouple to monitor the sample temperature and a heating controller to keep the temperature constant or to change it in a pre-determined fashion complete the essentials.

TGA is an extremely robust technique, in that the process is fundamentally simple, and the apparatus is similarly uncomplicated, making it difficult (although not completely impossible) to use the apparatus incorrectly. Sample preparation involves placing a powdered sample in an open cup of platinum or some other unreactive metal. The sample cup is attached to a balance arm, and the furnace is placed around the sample. The exact mechanical arrangement differs with the manufacturer, and the details are best obtained from the descriptive material that accompanies the equipment. The sample is usually heated while in a current of dry gas, typically nitrogen, but some TGA instruments are capable of operating in a vacuum or in various controlled atmospheres. The results of the experiment are recorded directly as weight loss vs. time or temperature. In contrast to the recording of weight loss, the interpretation of the data for kinetic purposes may be less straightforward, but that problem is common to DSC as well and will be discussed below.

Differential scanning calorimetry

Superficially, differential scanning calorimetry (DSC) and differential thermal analysis (DTA) appear to be similar. The apparent similarity lies in the fact that both measure the same types of reactions, produce data that are superficially similar, use the same terminology (e.g., endothermic reaction, exothermic reaction), and compare the thermal properties of a sample with those of a reference standard, but the physical principles behind the two techniques are completely different. Because the two techniques are often confused, the following discussion is considerably more detailed than that for TGA. An adequate understanding of what is being measured and how the measurements are made is absolutely essential in order to understand correctly the discussions of kinetics that follow.

In essence, DTA compares the temperature of a sample with that of a suitable reference material while both are being heated at the same rate. Any difference in temperature between the two is detected by thermocouples whose signal is proportional to ΔT (Figure 2). To maintain the sample and reference material under similar conditions, both are embedded in a block of material having a large thermal mass. In theory, the temperature change in the sample should be proportional to the enthalpy change, but this is not the case in practice because of the loss or gain of heat between the sample and the DTA apparatus. Various techniques have evolved to minimize this problem, but the technique remains of limited value for quantitative calorimetry.

The DSC apparatus is designed specifically to keep the sample and standard at the same (or nearly the same) temperature throughout the experiment. In the Perkin-Elmer DSC, the sample and reference holders are separate units, and each incorporates a resistance heater and a temperature sensor (a platinum resistance thermometer). While the sample and the reference are being heated linearly or in the isothermal mode, their temperatures are compared frequently, and any imbalance is compensated by increasing or decreasing the power going to the resistance heater in the sample cup (Figure 2). Compensation is rapid, because the thermal masses of the sample and standard cups are small, which gives this approach a clear advantage over DTA, in which the precise temperature of the sample is unknown during a reaction. In some instruments, the sample and standard are allowed to have a small, fixed difference in temperature, and the necessary power differential is measured. In practical terms, the difference between the two approaches is probably insignificant.

The differential power required to keep the two temperatures in the null condition is amplified and displayed. Thus, the signal that comes from the machine is in rate of energy change per unit time, or dH/dt, and has units of, e.g., mcal/s. From this come three useful quantities. The first is heat capacity, viz.,

$$C_p = dH/dT, \qquad (1)$$

where T is temperature and H is enthalpy. The abscissa of a graphical output is in

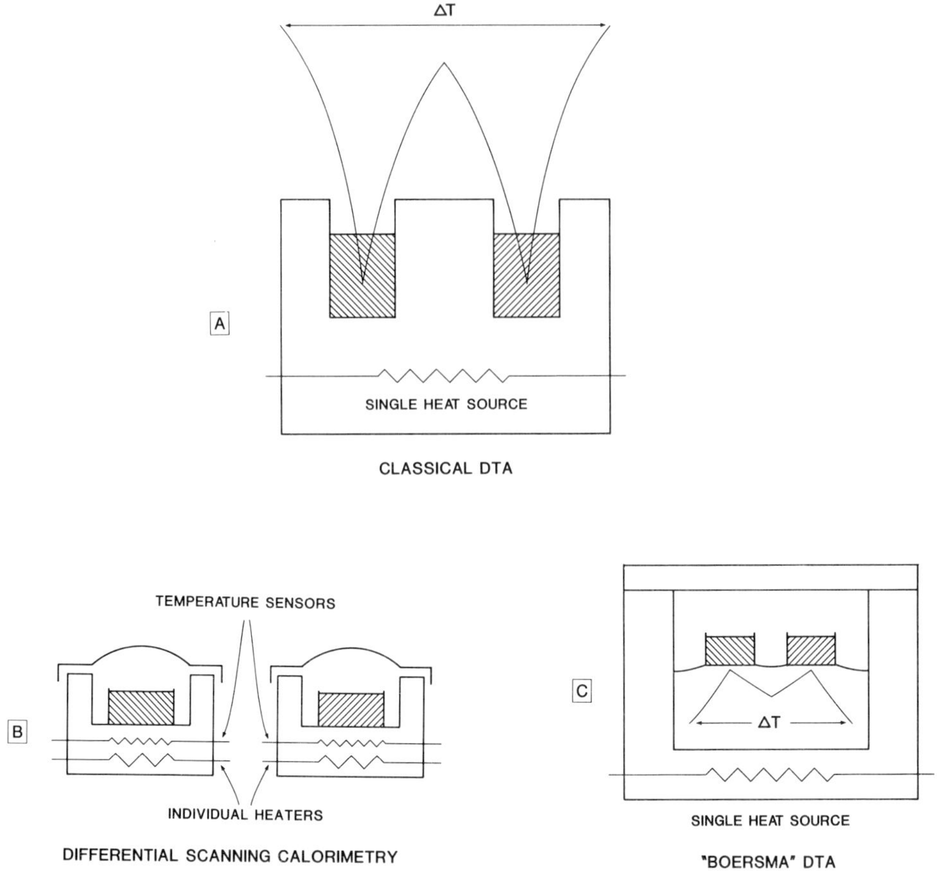

Figure 2. Diagrammatic representations of (a) classical differential thermal analysis (DTA) showing that the signal for a thermal event is given in terms of temperature difference; (b) a variant on DTA in which the thermal mass of the sample and reference containers is kept small; (c) differential scanning calorimetry (DSC), in which the thermal event is given by the difference in the power required to keep the sample and the reference at the same temperature.

units of time, which are related to T by the linear heating rate of the DSC. The relation can be rewritten as:

$$C_p = (dH/dt)/(dT/dt), \tag{2}$$

or

$$C_p = (dH/dt)\cdot(dt/dT), \tag{3}$$

which gives Eq. (1). Eq. (2) can be written in the form:

$$C_p = (\text{DSC signal})/(\text{DSC heating rate}). \tag{4}$$

The signal produced by the DSC in scanning from an initial to a final temperature is characteristic of the machine; this is very different from the output of a DTA. Figure 3 shows an idealized scan; the sections of the graphical record are the initial isothermal line at T_1 (1 in Figure 3), the onset of heating (2 in Figure 3), the attainment of equilibrium during linear heating (3 in Figure 3), the cessation of heating (4 in Figure 3), and the final isothermal at T_2. The displacement is taken as the distance from the signal at any temperature to the baseline at the same temperature. The baseline is an interpolated straight line, which must be drawn correctly. For T_1, the baseline begins just at the moment when the signal first departs from the isothermal; the T_2 end of the baseline is the intersection of the T_2 isothermal and a perpendicular dropped from the signal just before the end of the heating program causes the signal to deviate from the smooth scan. In the illustration, these two points are marked "program start" and "program end". The true signal is measured with reference to the empty holder scan, not the interpolated baseline.

Measurements. The ordinate scale can be converted to read directly in heat capacity units by dividing the full-scale sensitivity setting (in mcal/s) on the DSC machine (the "Range" setting on the Perkin-Elmer DSC-2 instrument) by the heating rate (in degrees/second). For initial measurements, this accuracy is often sufficient, but better results can be obtained by comparing the sample with a material of known C_p, as will be explained below.

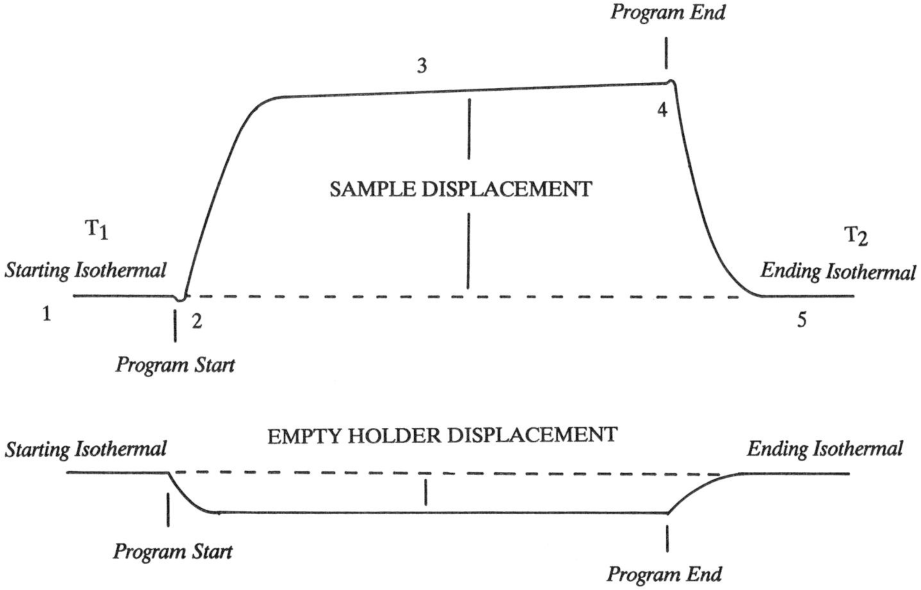

Figure 3. Idealized graphical record of a sample (here, aluminum oxide) kept at constant temperature (the "initial isotherm") shown by (1), at the onset of heating at a constant rate (event 2) during heating under equilibrium conditions (3), and at the moment of the cessation of heating (4).

The second interesting thermodynamic parameter that can be obtained from the DSC data is enthalpy, ΔH, for a reaction (e.g., deintercalation or dehydration) or a phase transition (e.g., melting). The determination of ΔH comes from

$$\Delta H = \int_{T_1}^{T_2} C_p(T)dt. \tag{5}$$

Because the pen displacement of the recorder along the ordinate can be represented in units of C_p the area under the DSC curve for a reaction is proportional to ΔH for the reaction. The area under the curve can be obtained using a planimeter, or the output of the DSC can be digitized (by hand or by means of an on-line computer) and integrated numerically. Little is said here about the determination of ΔH, because it is automatically determined during the study of the kinetics of a reaction (*vide infra*).

The final item of interest that follows from the DSC recording of an event is the ability to analyze the kinetics of the event. Assuming that the event takes place completely (i.e., 100% reaction) between temperatures T_1 and T_2, the area under the curve represents the entire process. Kinetics examines how the process develops with time (or temperature), that is, what fraction of the whole reaction has taken place at a given temperature lying between T_1 and T_2. This is a much more complicated process than the determination of enthalpy or specific heat and will be treated at length in a later section.

The exact manner in which the sample is loaded depends on the type of measurement to be made. Kinetics studies are far more complicated and prone to experimental difficulties than are measurements of C_p. Accordingly, sample manipulation for enthalpy and kinetics and for heat capacity will also be discussed in later sections.

Sample preparation and operation. The two cups that contain the sample and reference heaters and temperature sensors (Figure 2) normally have loose-fitting platinum caps. In turn, the block holding the two heaters is covered during a run with a loose fitting cylindrical piece of metal, which further isolates the sample and reference cups from the environment. This arrangement is necessary because the DSC measures very small changes in thermal properties. Both the sample and standard heaters are surrounded by slowly flowing dry nitrogen.

The sample or the reference standard is placed in a small (aluminum, gold, graphite, or other material) container or cup designed to fit into the sample or reference holder. The sample cup can be covered by a cap, and the whole assembly can then be crimped to seal the material in the metal cup. Whether the sample cup is covered or not is determined by the nature of the thermal reaction. Each situation will be treated below in appropriate sections.

The DSC is typically set up differently for different temperature regimes.

From ambient temperature to the upper limit (usually about 750°C), no extra precautions are necessary, the only caveat being that use of an aluminum cup is limited in temperature; for temperatures above the melting point of aluminum (600°C), gold is the material of choice.

Sub-ambient operation. Temperatures below ambient can be reached by attaching a cold sink to the head containing the sample and reference heaters. The cold sink can take many forms: a circulating water system with an ice bath, a refrigerator cooling block, or a cold finger attached to the block and immersed in a cold liquid (e.g., dry ice in acetone, liquid nitrogen). For these temperatures, frost accumulation on the head and other parts of the DSC becomes a problem, which can only be prevented by enclosing the top of the DSC with a dry box, constantly purged with dry nitrogen. For the lowest temperatures, using a liquid nitrogen heat sink, the gas flowing past the sample and reference heaters must be very dry helium. The use of helium changes the thermal parameters of the system, and the DSC must be recalibrated for temperature and response.

Working with liquid nitrogen is difficult, time consuming, and very frustrating. The problems are created by the enormous thermal gradient that exists between the sample and reference heaters and the liquid nitrogen. To record useful data, this gradient must remain unchanged throughout the experiments, which may, as will be shown later, consist of several scans of different samples over the same temperature regime. For this reason, it is best to work over very narrow temperature regions so that the last of the measurements is as close in time as possible to the first. This mode of operation, naturally, means that more scans must be made to cover the desired temperature region. For example, C_p measurements above ambient can be determined from scans that cover 50 to 100 K, with little trouble. At temperatures <200 K, using a liquid nitrogen bath, the scans should be 10 K in length. The author has often scanned temperature ranges of 2 K, acquiring only a single data point per scan. Very low temperature work is not a task to be taken lightly.

KINETICS OF DEINTERCALATION

The following is a brief introduction to kinetics, especially as applied to clay minerals containing intercalated species. It is not meant to be rigorous or complete, but should serve as an introduction to an extremely interesting aspect of reaction kinetics in general.

General

Reactions take place if a system becomes unstable, and the constituents of the system rearrange themselves so as to regain equilibrium. This rearrangement can take many forms, with changes of phase, decompositions, chemical reactions being a few. If the constituents are all in a common phase, e.g., an aqueous solution, the system is termed *homogeneous,* whereas if more than one phase exists, e.g., a solid in contact with a vapor, the system is said to be *inhomogeneous* or *heterogeneous.*

For homogeneous reactions, the rate at which the reaction is driven can be expressed in terms of a rate law having the general form $r = k[A]^n$, where k is the rate constant, A is the concentration of one of the reactants, and n is an integer or fraction, termed the order of the reaction. For homogeneous reactions, the order yields information about the nature of the reaction and may even reveal details about the mechanisms of the reaction.

Such is not the case for inhomogeneous reactions. Consider the following reaction:

$$A_{solid} = B_{solid} + C_{vapor}. \tag{6}$$

For this reaction, neither the solid nor the vapor can be easily discussed in terms of concentration. Also of importance is the fact that the system is not easily held in equilibrium, particularly during an experiment in which the conditions are changing. Clearly, the deintercalation of organic molecules or water from a clay mineral falls into this category. Therefore, the results of a kinetic analysis should be interpreted with these difficulties in mind.

Inhomogeneous rate laws

The mechanisms whereby a solid decomposes are varied, and this variety has led to the use of a fairly large number of kinetics laws. The complexity of the processes involved precludes knowing in advance of the experiment which of these kinetics laws will correctly describe the system being investigated, a rather serious situation. What must be done is to take the results of the experiment and see which of the laws provides a "best" fit to the data. As shown below, this procedure is seldom an easy task.

Descriptions of kinetics laws can be found in many standard texts on reaction rates. One of the more lucid discussions, and one that is particularly aimed at differential scanning calorimetric measurements, is the article by Guarini and Spinicci (1972). Their mathematical notation is often confusing, so the following derivations are more detailed than would normally be appropriate to avoid the confusion that might occur upon first reading their paper. The forms of the rate laws are listed in Table 1.

A kinetics law describes the rate at which a reaction takes place with time. The evolution of the experiment is described in terms of the fraction of the original sample that has reacted (termed, α) at a given time; the rate is given by $d\alpha/dt$. The transformation of a kinetics law into a form suitable for the DSC, or for any other thermal analytical technique, involves recasting the law into the form $d\alpha/dt =$

Table 1. List of commonly used kinetics rate laws and their mathematical formulations (Guarini and Spinicci, 1972)[1].

Law	Form
Contracting circle	$1 - (1 - \alpha)^{1/2} = kt$
Contracting sphere	$1 - (1 - \alpha)^{1/3} = kt$
Avrami-Erofeev	$-\ln(1 - \alpha) = (kt)^n$
Power	$\alpha = kt^n$
First order	$-\ln(1 - \alpha) = kt$
Prout-Tompkins	$(\ln \alpha)/(1 - \alpha) = kt + c$

[1]k = rate constant, t = time, α = fraction of sample that has reacted at time t. In a dynamic experiment in which heating rate is linear, t and temperature, T, are directly related; whereas for an isothermal experiment, T, of course, is constant and independent of t.

$f(k,\alpha)$, relating $d\alpha/dt$ to the parameters generated by the experiment (e.g., DSC signal, amplification setting of the DSC, sample weight) and solving for k. Two examples are presented below, one simple and the other a bit more complex.

Contracting circle. For the contracting circle model (see Young, 1966, for a derivation), the rate law is:

$$kt = 1 - (1 - \alpha)^{1/2}, \tag{7}$$

which gives, upon rearrangement,

$$(1 - \alpha) = (1 - kt)^2. \tag{8}$$

and

$$\alpha = 1 - (1 - kt)^2. \tag{9}$$

Differentiating with respect to t gives

$$d\alpha/dt = 2k(1 - kt), \tag{10}$$

and, going back to the original definition of the rate law (Eq. (7),

$$d\alpha/dt = 2k(1 - a)^{½}, \tag{11}$$

which is the result desired.

Avrami-Erofeev law. A more complicated example is furnished by the Avrami-Erofeev law. This kinetics law exists in various forms, but the following form is suited for the types of reactions being discussed:

$$-\ln(1 - \alpha) = (kt)^n. \tag{12}$$

Taking the exponential of both sides gives:

$$(1 - \alpha) = e^{-(kt)^n}, \tag{13}$$

and rearranging and differentiating gives:

$$d\alpha/dt = -nk(kt)^{n-1}e^{-(kt)^n}. \tag{14}$$

The substitution $\exp[-(kt)^n] = (1 - \alpha)$ gives:

$$d\alpha/dt = -nk(1 - \alpha)(kt)^{n-1}. \tag{15}$$

The exponential can be simplified by multiplying numerator and denominator by kt to give:

$$d\alpha/dt = -[nk(1 - \alpha)(kt)^n]/(kt). \tag{16}$$

The $(kt)^n$ factor in the numerator is equal to $-\ln(1 - \alpha)$; hence, the expression is now:

$$d\alpha/dt = [nk(1 - \alpha)\ln(1 - \alpha)]/[(-\ln(1 - \alpha)]^{1/n}, \tag{17}$$

which is the desired result. Similar treatment will transform any of the rate laws in Table 1 into the proper form of $d\alpha/dt = f(k,\alpha)$.

Differential scanning calorimetry

The next step is to relate the experimentally observed parameters from a DSC desorption run to the quantities specified by the rate law in question. Thomas and Clarke (1968) showed that for DSC, the rate of decomposition with time is given by:

$$d\alpha/dt = (1/\Delta H)(dH/dt), \tag{18}$$

where ΔH is proportional to the integrated area under the DSC curve, and dH/dt is proportional to the pen displacement at time t. These two quantities, the area of the curve (especially the area swept out at time t) and the instantaneous pen displacement, are the fundamental measurements needed for the calculation of the activation energy. The machine settings that give the proportionality constants are given by:

$$\Delta H = (A/S)(R/W), \tag{19}$$

where R is the sensitivity in terms of the full-scale deflection of the chart pen. R has units of mcal/s. A is the total area under the curve of dH/dt for the reaction (Figure 4), S is the chart speed in cm/s, and W is the chart width in cm.

As mentioned above, the instantaneous pen displacement, h_t, is proportional to the time rate of change of enthalpy:

$$dH/dt = h_t(R/W), \tag{20}$$

which, when combined with Eq. (19) gives:

$$d\alpha/dt = h_t(S/A). \tag{21}$$

Finally, $(1 - \alpha)$ and the quantities measured from the experiment are related by:

$$(1 - \alpha) = B_t/A, \tag{22}$$

where B_t is the area under the reaction curve that remains at time t, and A is the total area under the curve (Figure 4).

Two relations are thus established, one describing the rate law, the other involving only the quantities measured for the DSC run, are equal to the same thing, $d\alpha/dt$. Setting the rate law and the experimental relations equal to each other gives, for the contracting circle law:

$$h_tS/A = 2k(1 - \alpha)^{1/2}. \tag{23}$$

Substituting in the relations $(1 - \alpha) = B_t/A$ and solving for k gives:

$$k = (h_tS/2B_t)(B_t/A)^{1/2}. \tag{24}$$

The treatment of the Avrami-Erofeev equation follows essentially the same path. Here, the measured quantities and the rate law are related as:

$$h_tS/A = [nk(1 - \alpha)(\ln(1 - \alpha)]/[(-\ln(1 - \alpha)]^{1/n}. \tag{25}$$

Going back to the original statement of the law, $-\ln(1 - \alpha) = -\ln(B_t/A)$, which is the same as ln(A/B). Substituting this into Eq. (25) gives:

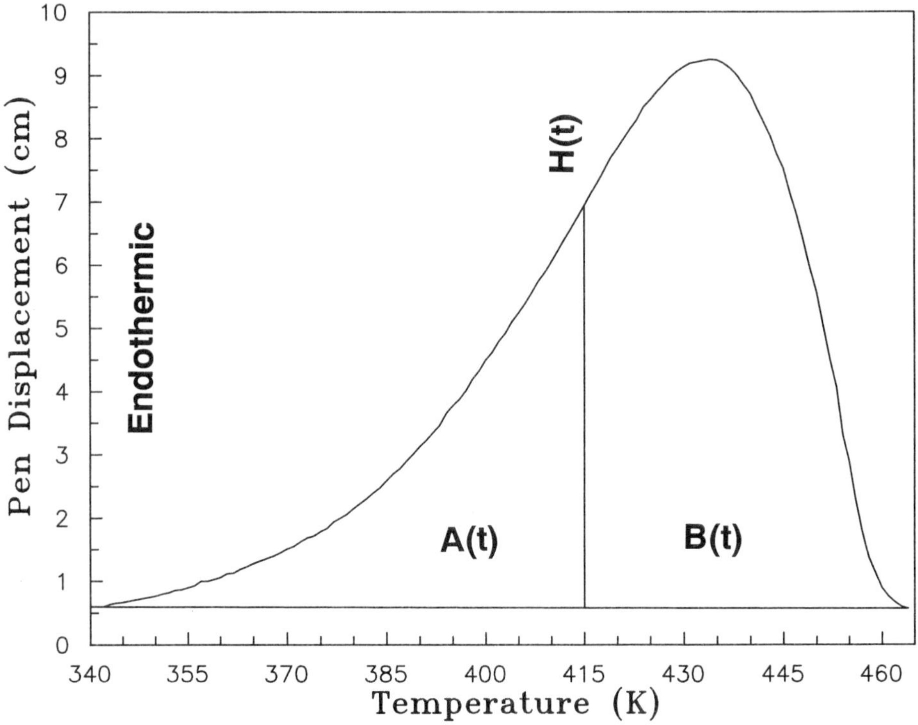

Figure 4. Part of a graphical record of the decomposition of Georgia kaolinite (KGa-1) intercalated with N-methyl formamide. Areas blocked out indicate the fraction of the sample that has already reacted (A_t), the fraction still unreacted (B_t) and the pen displacement (h_t) at time t. Time and temperature are linearly related by the instrument settings. Scale of y-axis is in strip-chart units. Straight line connecting onset and finish of the reaction is the baseline.

$$h_tS/A = [nk \ln(A/B)(B_t/A)]/\ln(A/B_t)^{1/n} \tag{26}$$

and

$$k = (h_tS/A)[\ln(A/Bt)^{1/n}]/[n \ln(A/B_t)(B_t/A)]. \tag{27}$$

Activation energy

The final relation gives the rate constant, k, in terms of areas, B_t and A, along with the pen displacement, h_t, at time t. Thus, k can be determined as a function of t (which is related to the temperature, T) for a series of points chosen along the x-axis of the recorder chart paper. The activation energy for the desorption is given by plotting the k values against 1000/T and constructing a regression line through the points. This is the Arrhenius plot, the slope of which gives the activation energy. The equation is:

$$k = Ae^{-Ea/RT}, \tag{28}$$

where A is the so-called "pre-exponential factor" or "Arrhenius factor", E_a is the activation energy, R is the gas constant, and T is the temperature. The units of E_a and RT are the same, energy/mole. For energy expressed in calories, R has a value of 1.987 cal/(mole K) or 8.314 J/(mole K). E_a can be determined graphically by taking the logarithm of each side of the equation, yielding:

$$\ln k = \ln A - (E_a/R). \tag{29}$$

This can be written in the form:

$$\ln k = \ln A - (E_a/R)(1/T), \tag{30}$$

which is a straight line relation with the y-axis in units of ln k, the x-axis in units of 1/T (or, as is often done, 1000/T), and a y-intercept of ln A. E_a is the slope of the straight line. The calculation of the activation energy is shown in Figure 5.

Determining the rate law

A general description of rate laws that have been used to study thermally driven decompositions or other such reactions was presented above, however, nothing was said about which law should be used to analyze the rate data produced by an experiment. The rate law can be chosen in two ways: one way is to try all laws and pick the one that gives the best fit to the data; the other is to conduct an auxiliary experiment that might suggest which is the appropriate law. The following discussion indicates how isothermal experiments might be used to indicate which rate law is best.

Consider an experiment in which a sample (e.g., an intercalated kaolinite) is rapidly heated to a temperature above that at which it becomes unstable and deintercalates. If the reaction is followed as a function of time, the manner by which α/t evolves with increasing t (k and T are now fixed) can be examined (Guarini and Spinicci, 1972). The purpose of this procedure can be seen by examining a specific example—the contracting circle law (*vide supra*), which is described by:

$$kt = 1 - (1 - \alpha)^{1/2}. \tag{31}$$

Rearranging to put "α" on the left of the equals sign gives:

$$\alpha = 2kt - k^2t^2, \tag{32}$$

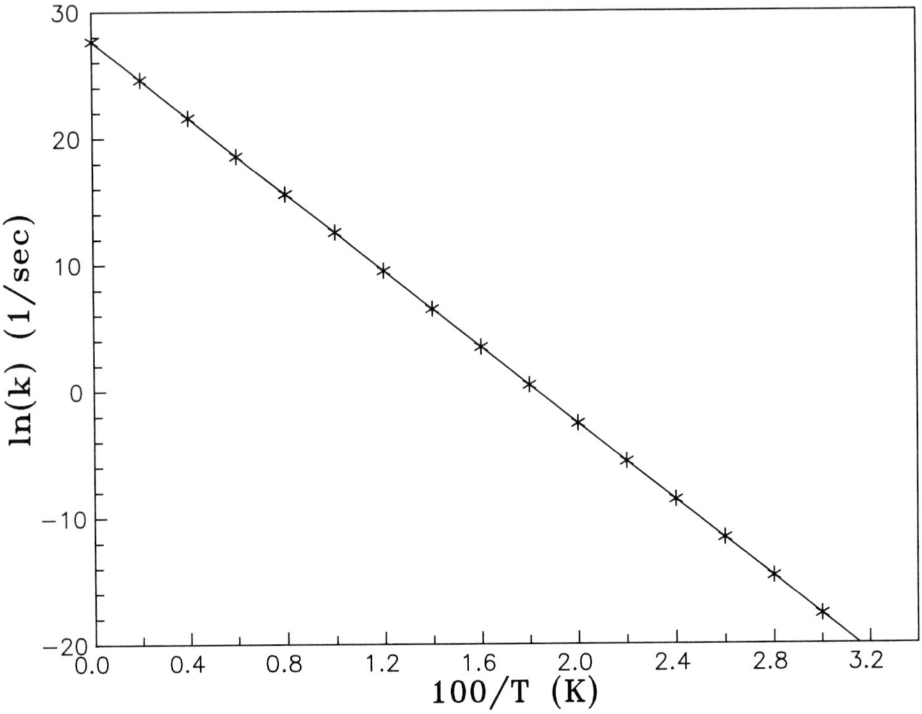

Figure 5. Arrhenius plot derived from a differential scanning calorimetric analysis of a decomposing material. Each of the squares is a data point representing the rate constant, k, determined for a specific rate law. If the rate law is applicable, the ln k values will fall on a straight line, which is extrapolated to 1000/T = 0 to give the pre-exponential constant of the Arrhenius equation. Slope of the line gives the activation energy for the reaction. See text for calculation.

and dividing through by t gives:

$$\alpha/t = 2k - k^2t. \quad (33)$$

This relation has the form of a straight line with a negative slope. Similar treatment of the Avrami-Erofeev law gives:

$$\alpha/t = k^2t - (k^4t^3)/2, \quad (34)$$

which is a function passing through the origin (ignoring the question of what happens to α/t if both α and t approach zero), passing through a maximum, and then declining. The power law with an exponent of 2 gives:

$$\alpha/t = kt, \quad (35)$$

a straight line through the origin. The contracting sphere gives:

$$\alpha/t = 3k - 3k^2t + k^3t^2, \tag{36}$$

which is seen to be a cubic equation with a concave upward shape ($d^2(\alpha/t)/dt^2 > 0$). Examples of each of these are shown in Figure 6.

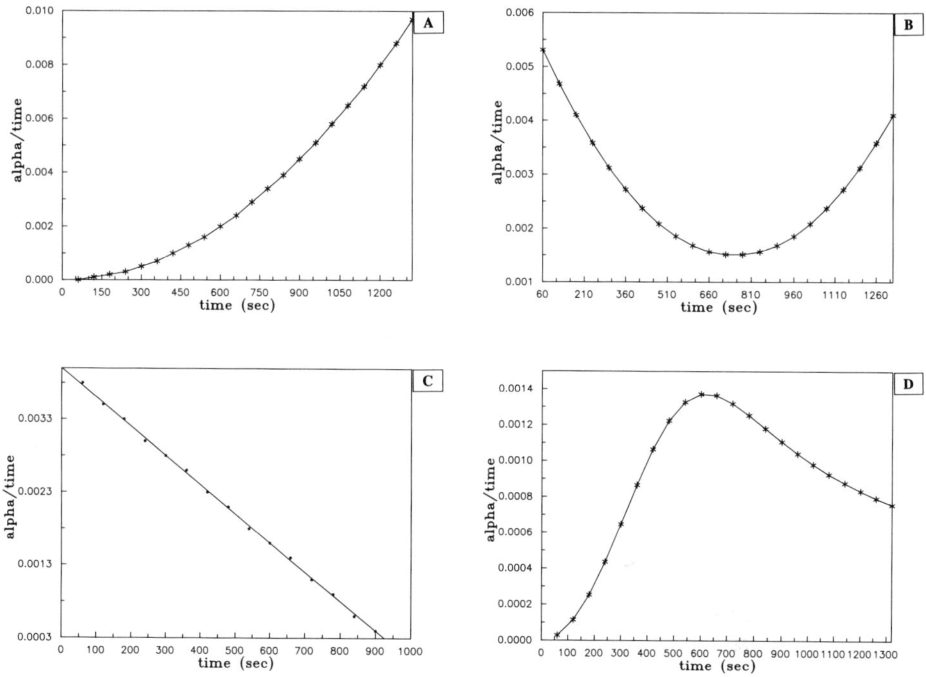

Figure 6. Weigh loss of a decomposing solid kept at constant temperature above the onset of decomposition. Ratio of fraction decomposed to the time after the start of the reaction is plotted (y-axis) vs. time. Reaction is for (A) a power law having an exponent of 2 (see text for description of this law); (B) contracting sphere law; (C) contracting circle law; (D) Avrami-Erofeev law.

In principle, which rate law is best should be forthcoming from a plot of the isothermal data as α/t vs. t. This technique seems not to have been widely used, but an example is found in Adams and Waltl (1980).

EXAMPLES: DEINTERCALATION

To illustrate the experimental set-up and the data reduction, a DSC run of well-crystallized kaolinite from Georgia (from the Source Clays Repository of The Clay Minerals Society), which had been intercalated with N-methyl formamide

(NMF) was used. Details of how to form such an intercalate are available in several articles (see Theng, 1974, for a summary).

Experimental conditions

Normally, if a sample is to be examined in a DSC, it is placed in an aluminum or gold sample pan (depending on the upper temperature limit), about 6 mm in diameter and 2 mm deep. Samples of clay minerals typically weigh < 15 mg. The open sample holder is then covered with a disc of the same metal, and, if desired, the two can be crimped together. This is the standard procedure described in the operating directions supplied with the machine; however, this is *not* the way to study such reactions as the deintercalation of kaolinite:NMF, the reason being that the desorbed organic material (or water, as the case may be) should be able to leave the sample as freely and as rapidly as possible. Thus, the sample container must be open to the atmosphere.

Normally the aluminum or gold sample holder is placed in the heating and sensing cup of the DSC, and the cup in turn is covered by a movable plate which has an O-ring to seal the entire assembly. Again, this procedure should not be followed for desorption reactions. Because the sample and reference should be as closely matched as possible, the reference sample holder should contain about the same quantity of material having a similar heat capacity as that of the sample.

Before the experiment is begun, the metal head, in which the sample and reference heating/sensing cups are situated, is capped by a pivoting metal block. This block should be replaced by a similar block that is vented. The problem with such an arrangement is that the sample and reference materials do not have the same thermal emissivity, which results in a nonlinear baseline. Not only is the baseline nonlinear, but the degree and even direction of nonlinearity are unpredictable a priori. As a result, the DSC curve may exceed the width of the chart paper and the run is wasted. This possibility can be reduced by making the ordinate less sensitive, but this results in a small signal, which is difficult to measure accurately. This problem is probably the single most vexing aspect of these measurements. Recording the pen displacement as a voltage (using, e.g., an on-line computer) may reduce this problem, but the author has no experience with this.

The simplest experiment requires a single run with the sample. The baseline against which the signal is measured can be taken as a straight line between what appears to be the start and finish of the reaction. Usually the end of the reaction is abrupt; it is accompanied by a rapidly declining signal and is rather easily identified. In contrast, the onset of deintercalation is often very gradual and more difficult to spot accurately. This requires some practice. A better baseline can be found by rerunning the same sample and reference through the same temperature range. The clay is now inert and should show the shape of the true baseline, which is often nonlinear.

The data to be used for the kinetics and enthalpy determination consist of pairs of readings, a temperature and a pen displacement, taken at regular intervals along the abscissa. The curve for the kaolinite:NMF intercalate is shown in Figure 4. The pen displacements can be obtained by careful measurement with a ruler or by recording the data directly with an on-line computer. The data in question were measured by hand at one-degree intervals. The machine settings are shown in Table 2.

Table 2. Settings for differential scanning calorimetry (DSC)[1].

Setting	Value for DSC	Value for computation	Name used in spread sheet
Heating rate	1°/min	0.167°/s	HEAT
Chart speed	20 mm/min	0.033 cm/s	CHART
Range[2]	5 mcal/s	0.2147 mcal/s/cm	RANGE
Sample weight	14.44 mg	1444 mg	SAMPLE
DSC constant	0.967	0.967	FACTOR

[1]Second column = as they appear in the machine; third column = as they are used in the computations; fourth column = symbolic name used in computations illustrated in Table 4.
[2]Full scale is 23 cm.

All the quantities have been described above, except the DSC constant. This constant is a multiplication factor obtained by measuring the enthalpy for a standard transition (e.g., the melting of high-purity indium) and comparing it with the known value. The constant compensates for the fact that the RANGE setting may be slightly different from the value indicated on the machine. Details can be found in the manual supplied with the DSC.

Computations

The computations described in the following sections are tedious rather than difficult. They can be easily programmed for a small digital computer, but the availability of very powerful spread-sheet programs on personal computers obviates any necessity for programming. Further, the mathematical commands used in spread sheets are readily understandable, and presenting the computations in a spread-sheet format is a useful pedagogical exercise. All references below are to the Lotus 1-2-3 spread sheet, although the commands are readily adapted to others.

Table 3 gives the Lotus 1-2-3 formulae that are used to calculate the values in the indicated columns of the spread sheet (Table 4); they are based on or use the contracting circle model. Note that cells H1 and I1 are blank; these values (the constant rate, k, and its logarithm) are undefined at the beginning of the reaction. Several quantities must be calculated to perform the manipulations in the table. The general steps in the calculation of the enthalpy for the reaction and subsequent analysis of the kinetics are:

1. Subtract the baseline from the pen displacements.
2. Compute the area for each increment in temperature using Simpson's rule (see Press *et al.*, 1986, Chapter 4, for a discussion of this and other methods of numerical integration).
3. Form a running sum of the area (the reaction which has already occurred) as tem perature increases = At.
4. Form a running sum of the area (the reaction which has not occurred) as a function of temperature = Bt.
5. Determine the entire area = A.
6. Calculate k according to the rate law chosen.
7. Calculate 1000/T for each k.
8. Do the Arrhenius plot and determine E for the reaction.

Table 3. Formulae used to compute contents of row 10 in each of the columns in the spread sheet in Table 4.

Column	Formula
C	B1O-(SLOPE*A10+INTERCEPT)
D	HEAT*(A10-A9)*(C10+C9)/2
E	D10+E9
F	AREA-E10
G	F10/AREA
H	@SQRT(G10)*(C10*CHART)/(2*F10)
I	@LN(H10)
J	1000/A10

[1]Names are symbolic and are defined in Table 2. Row 10 was chosen arbitrarily for illustration. The same formulae, with appropriate changes in row number, apply to any given row.

Intermediate quantities must also be computed. These are:

SLOPE	(B125-B1)/(A125-A1)
INTERCEPT	+B125-SLOPE*A125
AREA	@SUM(D1..D125)

Although unnecessay for the kinetic determination, the enthalpy change for the reaction can be easily found from the quantities in the spread sheet along with the DSC settings (*vide supra*) and the quantities above

ENTHALPY = (FACTOR*AREA*RANGE*)/(SAMPLE*CHART)

Table 4 shows approximately how the Lotus spread sheet should appear. Not shown are the constants that describe the DSC settings and the calculation of the baseline and area under the curve. These exist in the spread sheet in specific cells, and are called either by an address or by an assigned symbolic name. In the following discussion, symbolic names are used.

HEAT CAPACITY MEASUREMENT

Heat capacity, C_p, is a fundamental thermodynamic property, being related to the vibrational modes of the atoms in the solid. Thus, it is influenced by the degree of disorder in the system. This relationship is of some interest for materials in which part of the system undergoes a change from a high degree of order to a lower degree, as, for example, if a molecular species that is rigidly held in a crystalline environment gains rotational and/or translational freedom as the temperature is raised. Such a transition should be reflected in C_p.

The measurement of C_p with a DSC was alluded to above and is elaborated on here. Earlier, pen displacement in the y direction was shown to be proportional to dH/dt, and using the heating rate, dT/dt, pen displacement is converted into C_p units. The relation is:

$$Cp = \frac{(\text{Range/Width})}{(\text{Scan Rate/60})} \times \frac{\text{Sample Disp} + \text{Blank Disp}}{\text{Sample Wt. (mg)}} .$$

Here, the pen displacement is taken with respect to both the interpolated baseline and the position of the pen when the sample and reference pans are empty (blank run). If the pen for the blank run is on the other side of the baseline (as in Figure 3), the two displacements are added, otherwise the blank is subtracted from the sample displacement.

Table 4. Computation of contracting circle kinetics as done in a Lotus 1-2-3 spead-sheet format. [Cell addresses are designated by column (letter) and row (number). Formulae for computations are given in Table 3.]

	A	B	C	D	E	F	G	H	I	J
1	340	0.60	0.000000	0.000000	0.000000	88.53300	1.000000	-	-	2.93
2	341	0.60	0.000241	0.000024	0.000024	88.53297	0.999999	4.555E-08	-16.9045	2.93
3	342	0.60	0.000483	0.000072	0.000096	88.53290	0.999998	9.109E-08	-16.2114	2.92
4	343	0.64	0.040725	0.004120	0.004217	88.52878	0.999952	7.667E-06	-11.7785	2.92
5	344	0.66	0.060967	0.010169	0.014387	88.51861	0.999837	1.148E-05	-11.3750	2.91
6	345	0.67	0.071209	0.013217	0.027604	88.50539	0.999688	1.341E-05	-11.2196	2.90
7	346	0.69	0.091451	0.016266	0.043870	88.48912	0.999504	1.722E-05	-10.9694	2.89
8	347	0.71	0.111693	0.020314	0.064185	88.46881	0.999275	2.103E-05	-10.7693	2.88
9	348	0.73	0.131935	0.024362	0.088548	88.44445	0.998999	2.485E-05	-10.6026	2.87
10	349	0.75	0.152177	0.028411	0.116959	88.41604	0.998678	2.867E-05	-10.4597	2.87
11	350	0.77	0.172419	0.032459	0.149419	88.38358	0.998312	3.249E-05	-10.3347	2.86
12	351	0.80	0.202661	0.037508	0.186927	88.34607	0.997888	3.819E-05	-10.1728	2.85
13	352	0.82	0.222903	0.042556	0.229483	88.30351	0.997407	4.202E-05	-10.0774	2.84
14	353	0.86	0.263145	0.048604	0.278088	88.25491	0.996858	4.962E-05	-9.91119	2.83
15	354	0.87	0.273387	0.053653	0.331741	88.20125	0.996252	5.156E-05	-9.87270	2.82
16	355	0.90	0.303629	0.057701	0.389443	88.14355	0.995601	5.729E-05	-9.76746	2.82
17	356	0.93	0.333870	0.06375	0.453193	88.07980	0.994881	6.301E-05	-9.67215	2.81
18	357	1.00	0.404112	0.073798	0.526991	88.00600	0.994047	7.630E-05	-9.48079	2.80
19	358	1.00	0.404354	0.080846	0.607838	87.92516	0.993134	7.638E-05	-9.47973	2.79
20	359	1.03	0.434596	0.083895	0.691733	87.84126	0.992186	8.214E-05	-9.40713	2.79
21	360	1.06	0.464838	0.089943	0.781677	87.75132	0.991170	8.790E-05	-9.33935	2.78
22	361	1.12	0.525080	0.098991	0.880669	87.65233	0.990052	9.934E-05	-9.21692	2.77
23	362	1.13	0.535322	0.106040	0.986709	87.54629	0.988854	1.013E-04	-9.19700	2.76
24	363	1.19	0.595564	0.113088	1.099798	87.43320	0.987577	1.128E-04	-9.08971	2.75
25	364	1.23	0.635806	0.123137	1.222935	87.31006	0.986186	1.205E-04	-9.02362	2.75
26	365	1.28	0.686048	0.132185	1.355120	87.17787	0.984693	1.302E-04	-8.94681	2.74
27	366	1.32	0.726290	0.141233	1.496354	87.03664	0.983098	1.379E-04	-8.88900	2.73
28	367	1.36	0.766532	0.149282	1.645637	86.88736	0.981412	1.457E-04	-8.83421	2.72
29	368	1.40	0.806774	0.157330	1.802967	86.73003	0.979635	1.534E-04	-8.78214	2.72
30	369	1.45	0.857016	0.166379	1.969346	86.56365	0.977755	1.632E-04	-8.72077	2.71
31	370	1.51	0.917258	0.177427	2.146774	86.38622	0.975751	1.748E-04	-8.65181	2.70
32	371	1.55	0.957500	0.187475	2.334250	86.19875	0.973634	1.827E-04	-8.60778	2.70
33	372	1.63	1.037741	0.199524	2.533774	85.99922	0.971380	1.982E-04	-8.52615	2.69
34	373	1.69	1.097983	0.213572	2.747346	85.78565	0.968968	2.100E-04	-8.46848	2.68
35	374	1.72	1.128225	0.222620	2.969967	85.56303	0.966453	2.160E-04	-8.44001	2.67
36	375	1.79	1.198467	0.232669	3.202637	85.33036	0.963825	2.298E-04	-8.37825	2.67
37	376	1.84	1.248709	0.244717	3.447354	85.08564	0.961061	2.398E-04	-8.33575	2.66
38	377	1.94	1.348951	0.259766	3.707120	84.82587	0.958127	2.594E-04	-8.25700	2.65
39	378	2.00	1.409193	0.275814	3.982935	84.55006	0.955011	2.715E-04	-8.21168	2.65
40	379	2.06	1.469435	0.287862	4.270798	84.26220	0.951760	2.836E-04	-8.16812	2.64
41	380	2.15	1.559677	0.302911	4.573709	83.95929	0.948338	3.015E-04	-8.10671	2.63
42	381	2.23	1.639919	0.319959	4.893669	83.63933	0.944724	3.176E-04	-8.05464	2.62
43	382	2.31	1.720161	0.336008	5.229677	83.30332	0.940929	3.338E-04	-8.00485	2.62
44	383	2.40	1.810403	0.353056	5.582733	82.95026	0.936941	3.521E-04	-7.95160	2.61
45	384	2.48	1.890645	0.370104	5.952838	82.58016	0.932761	3.685E-04	-7.90599	2.60
46	385	2.59	2.000887	0.389153	6.341991	82.19100	0.928365	3.909E-04	-7.84696	2.60
47	386	2.70	2.111129	0.411201	6.753193	81.77980	0.923721	4.135E-04	-7.79082	2.59
48	387	2.78	2.191370	0.43025	7.183443	81.34955	0.918861	4.304E-04	-7.75088	2.58
49	388	2.90	2.311612	0.450298	7.633741	80.89925	0.913775	4.552E-04	-7.69468	2.58
50	389	3.01	2.421854	0.473346	8.107088	80.42591	0.908428	4.784E-04	-7.64516	2.57
	T (K)	Disp. cm	Corr. Disp. cm	Incr. Area cm^2	A_t	B_t	B_t/A_t	k	ln(k)	1000/T

Table 4 (continued).

	A	B	C	D	E	F	G	H	I	J
51	390	3.13	2.542096	0.496395	8.603483	79.92951	0.902821	5.037E-04	-7.59361	2.56
52	391	3.23	2.642338	0.518443	9.121927	79.41107	0.896965	5.252E-04	-7.55168	2.56
53	392	3.35	2.762580	0.540491	9.662419	78.87058	0.890860	5.510E-04	-7.50377	2.55
54	393	3.46	2.872822	0.563540	10.22595	78.30704	0.884495	5.750E-04	-7.46105	2.54
55	394	3.65	3.063064	0.593588	10.81954	77.71345	0.877790	6.155E-04	-7.39313	2.54
56	395	3.78	3.193306	0.625637	11.44518	77.08781	0.870724	6.442E-04	-7.34744	2.53
57	396	3.88	3.293548	0.648685	12.09387	76.43912	0.863397	6.673E-04	-7.31231	2.53
58	397	4.02	3.433790	0.672733	12.76660	75.76639	0.855798	6.988E-04	-7.26619	2.52
59	398	4.20	3.614032	0.704782	13.47138	75.06161	0.847837	7.389E-04	-7.21036	2.51
60	399	4.33	3.744274	0.735830	14.20721	74.32578	0.839526	7.693E-04	-7.17003	2.51
61	400	4.50	3.914516	0.765879	14.97309	73.55990	0.830875	8.085E-04	-7.12039	2.50
62	401	4.62	4.034758	0.794927	15.76802	72.76497	0.821896	8.378E-04	-7.08470	2.49
63	402	4.79	4.205000	0.823975	16.59200	71.94100	0.812589	8.782E-04	-7.03768	2.49
64	403	4.92	4.335241	0.854024	17.44602	71.08697	0.802943	9.108E-04	-7.00120	2.48
65	404	5.10	4.515483	0.885072	18.33109	70.20190	0.792946	9.546E-04	-6.95420	2.48
66	405	5.24	4.655725	0.917120	19.24821	69.28478	0.782587	9.908E-04	-6.91704	2.47
67	406	5.39	4.805967	0.946169	20.19438	68.33861	0.771899	1.030E-03	-6.87841	2.46
68	407	5.56	4.976209	0.978217	21.17260	67.36039	0.760850	1.074E-03	-6.83639	2.46
69	408	5.74	5.156451	1.013266	22.18587	66.34712	0.749405	1.121E-03	-6.79323	2.45
70	409	5.88	5.296693	1.045314	23.23118	65.30181	0.737598	1.161E-03	-6.75845	2.44
71	410	6.05	5.466935	1.076362	24.30754	64.22545	0.725440	1.208E-03	-6.71851	2.44
72	411	6.23	5.647177	1.111411	25.41895	63.11404	0.712887	1.259E-03	-6.67734	2.43
73	412	6.41	5.827419	1.147459	26.56641	61.96658	0.699926	1.311E-03	-6.63675	2.43
74	413	6.58	5.997661	1.182508	27.74892	60.78407	0.686569	1.363E-03	-6.59832	2.42
75	414	6.76	6.177903	1.217556	28.96648	59.56651	0.672817	1.418E-03	-6.55860	2.42
76	415	6.94	6.358145	1.253604	30.22008	58.31291	0.658657	1.475E-03	-6.51920	2.41
77	416	7.13	6.548387	1.290653	31.51074	57.02225	0.644079	1.536E-03	-6.47853	2.40
78	417	7.30	6.718629	1.326701	32.83744	55.69555	0.629093	1.595E-03	-6.44109	2.40
79	418	7.49	6.908870	1.36275	34.20019	54.33280	0.613701	1.660E-03	-6.40079	2.39
80	419	7.69	7.109112	1.401798	35.60199	52.93100	0.597867	1.731E-03	-6.35914	2.39
81	420	7.84	7.259354	1.436846	37.03883	51.49416	0.581638	1.792E-03	-6.32447	2.38
82	421	8.00	7.419596	1.467895	38.50673	50.02626	0.565057	1.858E-03	-6.28818	2.38
83	422	8.16	7.579838	1.499943	40.00667	48.52632	0.548115	1.927E-03	-6.25159	2.37
84	423	8.30	7.720080	1.529991	41.53666	46.99633	0.530834	1.995E-03	-6.21724	2.36
85	424	8.50	7.920322	1.564040	43.10070	45.43229	0.513167	2.081E-03	-6.17471	2.36
86	425	8.61	8.030564	1.595088	44.69579	43.83720	0.495150	2.148E-03	-6.14301	2.35
87	426	8.74	8.160806	1.619137	46.31493	42.21806	0.476862	2.225E-03	-6.10811	2.35
88	427	8.85	8.271048	1.643185	47.95812	40.57487	0.458302	2.300E-03	-6.07484	2.34
89	428	8.97	8.391290	1.666233	49.62435	38.90864	0.439481	2.383E-03	-6.03944	2.34
90	429	9.06	8.481532	1.687282	51.31163	37.22136	0.420423	2.462E-03	-6.00658	2.33
91	430	9.13	8.551774	1.703330	53.01496	35.51803	0.401184	2.542E-03	-5.97491	2.33
92	431	9.19	8.612016	1.716379	54.73134	33.80165	0.381797	2.624E-03	-5.94312	2.32
93	432	9.21	8.632258	1.724427	56.45577	32.07722	0.362319	2.700E-03	-5.91459	2.31
94	433	9.23	8.652500	1.728475	58.18425	30.34875	0.342795	2.782E-03	-5.88456	2.31
95	434	9.25	8.672741	1.732524	59.91677	28.61622	0.323226	2.872E-03	-5.85283	2.30
96	435	9.24	8.662983	1.733572	61.65034	26.88265	0.303645	2.960E-03	-5.82271	2.30
97	436	9.19	8.613225	1.727620	63.37796	25.15503	0.284131	3.042E-03	-5.79526	2.29
98	437	9.11	8.533467	1.714669	65.09263	23.44036	0.264764	3.122E-03	-5.76926	2.29
99	438	9.00	8.423709	1.695717	66.78835	21.74464	0.245610	3.200E-03	-5.74466	2.28
100	439	8.85	8.273951	1.669766	68.45812	20.07487	0.226750	3.271E-03	-5.72265	2.28
	T (K)	Disp. cm	Corr. Disp. cm	Incr. Area cm^2	A_t	B_t	B_t/A_t	k	ln(k)	1000/T

Table 4. (continued).

	A	B	C	D	E	F	G	H	I	J
101	440	8.71	8.134193	1.640814	70.09893	18.43406	0.208216	3.356E-03	-5.69705	2.27
102	441	8.48	7.904435	1.603862	71.70279	16.83020	0.190100	3.413E-03	-5.68019	2.27
103	442	8.27	7.694677	1.559911	73.26270	15.27029	0.172481	3.488E-03	-5.65845	2.26
104	443	8.05	7.474919	1.516959	74.77966	13.75333	0.155346	3.570E-03	-5.63511	2.26
105	444	7.76	7.185161	1.466008	76.24567	12.28732	0.138788	3.631E-03	-5.61829	2.25
106	445	7.53	6.955403	1.414056	77.65973	10.87326	0.122815	3.736E-03	-5.58966	2.25
107	446	7.14	6.565645	1.352104	79.01183	9.521161	0.107543	3.769E-03	-5.58093	2.24
108	447	6.80	6.225887	1.279153	80.29099	8.242008	0.093095	3.841E-03	-5.56193	2.24
109	448	6.40	5.826129	1.205201	81.49619	7.036806	0.079482	3.890E-03	-5.54925	2.23
110	449	5.96	5.386370	1.12125	82.61744	5.915556	0.066817	3.923E-03	-5.54095	2.23
111	450	5.57	4.996612	1.038298	83.65574	4.877258	0.055089	4.008E-03	-5.51956	2.22
112	451	5.03	4.456854	0.945346	84.60108	3.931911	0.044411	3.981E-03	-5.52615	2.22
113	452	4.51	3.937096	0.839395	85.44048	3.092516	0.034930	3.966E-03	-5.53008	2.21
114	453	4.07	3.497338	0.743443	86.18392	2.349072	0.026533	4.042E-03	-5.51104	2.21
115	454	3.31	2.737580	0.623491	86.80741	1.725580	0.019490	3.691E-03	-5.60173	2.20
116	455	2.89	2.317822	0.505540	87.31295	1.220040	0.013780	3.717E-03	-5.59484	2.20
117	456	2.29	1.718064	0.403588	87.71654	0.816451	0.009222	3.368E-03	-5.69344	2.19
118	457	1.79	1.218306	0.293637	88.01018	0.522814	0.005905	2.985E-03	-5.81430	2.19
119	458	1.38	0.808548	0.202685	88.21287	0.320129	0.003615	2.531E-03	-5.97903	2.18
120	459	1.13	0.558790	0.136733	88.34960	0.183395	0.002071	2.311E-03	-6.06995	2.18
121	460	0.89	0.319032	0.087782	88.43738	0.095612	0.001079	1.828E-03	-6.30477	2.17
122	461	0.76	0.189274	0.050830	88.48821	0.044782	0.000505	1.584E-03	-6.44761	2.17
123	462	0.67	0.099516	0.028879	88.51709	0.015903	0.000179	1.398E-03	-6.57285	2.16
124	463	0.60	0.029758	0.012927	88.53002	0.002975	0.000033	9.663E-04	-6.94206	2.16
125	464	0.57	0.000000	0.002975	88.53300	0.000000	0.000000	-	-	2.16
	T (K)	Disp. cm	Corr. Disp. cm	Incr. Area cm^2	A_t	B_t	B_t/A_t	k	ln(k)	1000/T

This method of determining the heat capacity of a sample is only approximate; a better method uses a standard material of known heat capacity. For high temperatures (above ambient), high purity aluminum oxide is used, whereas for low temperatures, benzoic acid is preferred. Values for the heat capacity of these materials can be found in Ginnings and Furukawa (1953). The method involves separate measurements: for the standards, the sample, and the sample and reference pans when empty (termed the blank). The technique was used by Lipsicas *et al.* (1986) to measure C_p for kaolinite. The computation is:

$$C_{p(sample)} = (W_{standard}/W_{sample})(h_{sample}/h_{standard})(C_{p(standard)}). \tag{38}$$

EXAMPLES: HEAT CAPACITY

Two topics are treated in this section. First, the determination of C_p for a single material or an intercalated clay mineral is outlined in a simple manner, which involves no major uncertainties, save for the usual caveats about careful experimental procedures and the use of a well-calibrated calorimeter. Second, a method is illustrated for determining C_p for the intercalated phase, i.e., that part of the C_p due to the molecules (or ionic species) that have been inserted between the clay layers.

Single material

The following is an example of the computations performed on scanning a sample (here, well-crystallized Georgia kaolinite) over a narrow temperature range (2°-5°C typically). As before, the computations are illustrated using a spread-sheet format in order to show the development of the procedure.

Experimental conditions. In contrast to the kinetics experiments, the standard conditions of sample preparation for C_p measurements are as described in the operating instructions supplied with the DSC; i.e., the sample and standard are placed in the metal cups (typically aluminum) and covered with a metal disc. Perkin-Elmer supplies standards for heat capacity measurements in the form of small discs of high-purity aluminum oxide. These discs are fairly expensive and can be replaced by high-purity aluminum oxide powder, which is available from several sources (e.g., Aldrich Chemical Company, part number 20,254-1, Gold Label aluminum oxide). If an alumina disc is used, the standard container should be uncrimped, but good practice is to crimp the sample into its cup. In addition, the sample and standard should be dried in an oven (at ≃150°C) to remove any surface-adsorbed water, which can influence C_p.

At least three runs over the same temperature region are necessary. The upper and lower temperature bounds should be <50°C apart (and <<50°C for subambient conditions). Longer runs can be made, but, in practice, the results are less reliable, probably because the thermal characteristics of the calorimeter change with time, the longer the run, the more likely that when the sample is replaced by the standard, the machine will respond differently to the programmed heating.

The order in which the three runs are made is of little consequence, but probably the best practice is to make all the measurements on a given sample in the same sequence. Scans of the sample, the standard, and the empty cup are made under identical conditions. The baseline is drawn for each according to the discussion in the section on differential scanning calorimetry. At convenient intervals, e.g., every 5° or 10°C, the pen displacements with respect to the baseline are measured for each of the three scans.

The DSC settings for the run under discussion are shown in Table 5. Note that the RANGE setting is smaller (greater amplification) than was used for the kinetics experiment. A large amplification produces a signal that is a more accurate measure of C_p. At the same time, as mentioned above, the risk is that part of the scan will go off scale, resulting in a failed run. This is less of a problem because the same sample can be re-run, which is impossible for the kinetics experiments.

Table 5. Differential scanning calorimetry (DSC) settings for heat capacity run described in Table 7.[1]

Setting	Value for DSC	Value of computation	Name used in spread sheet
Heating range	20°/min	0.333°/s	HEAT
Range[2]	2 mcal/s	0.0870 mcal/s/cm	RANGE
Sample wt.	12.72 mg	12.72 mg	SAMPLE
Standard wt. (benzoic acid)	14.6 mg	14.6 mg	STANDARD

[1]Values for machine and values used in computations are shown along with symbolic names used in spread-sheet computations.
[2]Full scale is 23 cm.

The scheme for the computations is as follows:

1. For both the sample and the standards, add (or subtract) the pen displacement for the blank.
2. As a check, directly compute C_p for the standard.
3. Calculate C_p for the sample by using the displacements of both sample and standard.

Note that the true pen displacement for the sample and standard is the distance from the pen position when the sample/standard container is empty, rather than the distance with respect to the baseline. The former is given by the blank run, and, if the blank displacement is on the opposite side of the baseline from that of the sample and the standard, the blank is added to each; otherwise the baseline is subtracted from the two runs.

Computations. In addition to the DSC settings, two other quantities are needed for the calculations: (1) the molecular weight of the sample, and (2) a constant that converts the strip chart ordinate into heat capacity units. These are:

MOLE	258.16
SCALE	((RANGE/23.)/HEAT)

The computation of C_p, is straightforward, with no intermediate calculations. Table 7 shows the spread sheet and Table 6 gives the formulae for each of the columns.

Table 6. Formulae used for heat capacity computations using data from Table 7.[1]

Column	Formula
E	+SCALE*(C1+B1)/STANDARD
F	[data from literature; see text]
G	100*(E1-F1)/F1
H	(STANDARD/SAMPLE)*((D1+81)/(C1+81))*F1
I	4.184*H1
J	MOLE*H1

[1]Symbolic names are defined in Table 2.

Intercalated material

A large difference should be expected in the environment around a molecule if it is intercalated in a clay mineral, compared with the same molecule in a bulk or pure state. This difference should be reflected in the C_p of the intercalated phase. The measurement of the C_p of the intercalated clay (as above) is easy, but to apportion the C_p between the clay fraction and the intercalate fraction is less obvious. The following is a description of how this can be done, and is followed by a brief description of the problems inherent in such a treatment.

An intercalated clay may be considered as two phases intimately admixed—the clay part and the intercalated material. Each has a heat capacity, which should be additive to give the total heat capacity of the intercalated clay. The relationship is (Cruz *et al.*, 1978):

$$C_p(\text{total}) = xC_p(\text{guest}) + (1 - x)C_p(\text{clay}), \tag{39}$$

where x is the weight fraction of guest and (1 - x) is the weight fraction of clay in the intercalated clay. In practice, C_p(clay) and C_p(total) are measured and C_p (guest) is computed. Although algebraically correct, the relation is not strictly valid (Sposito and Prost, 1982) in a thermodynamic sense, because the quantity C_p(clay) cannot be measured directly. The C_p(clay) is estimated, however, by measuring the heat capacity of the deintercalated clay. The problem, is that the C_p of the deintercalated clay is not the same as the heat capacity of the clay if the layers are separated by guest molecules. The technique has been used, nevertheless, to examine both water and adsorbed organic molecules.

Table 7. Data obtained from heat capacity measurements to illustrate the computations involved. [Data are presented as a spread sheet; cell addresses are designated by column (letter) and row (number). Formulae for computations that generated non-data columns are listed in Table 6.]

	A	B	C	D	E	F	G	H	I	J
1	190	4.28	7.03	4.44	0.2021	0.19327	4.6	0.1710	0.7156	44.15
2	195	3.82	8.01	5.22	0.2114	0.19732	7.1	0.1731	0.7241	44.68
3	200	3.50	8.60	5.79	0.2162	0.20137	7.4	0.1775	0.7425	45.81
4	205	3.36	9.04	6.19	0.2216	0.20550	7.8	0.1817	0.7601	46.90
5	210	3.30	9.38	6.48	0.2266	0.20963	8.1	0.1856	0.7765	47.91
6	215	3.29	9.64	6.69	0.2310	0.21383	8.0	0.1894	0.7926	48.91
7	220	3.35	9.85	6.89	0.2359	0.21803	8.2	0.1941	0.8123	50.12
8	225	3.48	10.03	6.98	0.2414	0.22232	8.6	0.1976	0.8266	51.00
9	230	3.63	10.11	7.02	0.2455	0.22660	8.3	0.2016	0.8435	52.04
10	235	3.74	10.18	6.99	0.2487	0.23095	7.7	0.2043	0.8549	52.75
11	240	3.95	10.22	6.98	0.2532	0.23529	7.6	0.2083	0.8716	53.78
12	245	4.17	10.24	6.92	0.2575	0.23968	7.4	0.2117	0.8858	54.66
13	250	4.43	10.26	6.88	0.2625	0.24407	7.5	0.2157	0.9024	55.68
14	255	4.67	10.20	6.83	0.2657	0.24851	6.9	0.2206	0.9230	56.95
15	260	4.87	10.12	6.74	0.2678	0.25294	5.9	0.2249	0.9408	58.05
16	261	4.91	10.10	6.74	0.2682	0.25383	5.7	0.2261	0.9461	58.38
17	262	4.97	10.10	6.73	0.2693	0.25473	5.7	0.2270	0.9497	58.60
18	263	4.99	10.10	6.72	0.2696	0.25562	5.5	0.2277	0.9526	58.78
19	264	5.02	10.06	6.69	0.2694	0.25652	5.0	0.2286	0.9566	59.02
20	265	5.08	10.05	6.65	0.2703	0.25741	5.0	0.2291	0.9584	59.13
21	266	5.13	10.03	6.62	0.2709	0.25830	4.9	0.2298	0.9615	59.32
22	267	5.13	10.02	6.61	0.2707	0.25920	4.4	0.2305	0.9646	59.52
23	268	5.21	9.99	6.56	0.2716	0.26009	4.4	0.2312	0.9672	59.68
24	269	5.26	9.97	6.52	0.2721	0.26099	4.3	0.2317	0.9694	59.82
25	270	5.30	9.92	6.45	0.2719	0.26188	3.8	0.2321	0.9709	59.91
26	271	5.34	9.91	6.41	0.2725	0.26278	3.7	0.2324	0.9723	59.99
27	272	5.40	9.90	6.38	0.2734	0.26368	3.7	0.2330	0.9750	60.16
28	273	5.46	9.85	6.31	0.2736	0.26458	3.4	0.2335	0.9768	60.27
29	274	5.52	9.82	6.24	0.2741	0.26548	3.2	0.2336	0.9774	60.31
30	275	5.63	9.73	6.21	0.2744	0.26638	3.0	0.2357	0.9861	60.84
	T (K)	Blank displ. cm	Standard displ. cm	Sample displ. cm	Standard Cp cal/g-K	Standard lit. value	Diff. %	Sample Cp cal/g-K	Sample Cp J/g-K	Sample Cp cal/moleK

Experimental conditions. The experimental conditions for measuring heat capacity of the intercalate are essentially the same as described above for a single material. The intercalated clay is sealed in the aluminum cup and scanned in the DSC over the temperature range of interest. The deintercalated clay material can be either the original intercalated clay or an intercalated sample that has been heated to drive off the adsorbed phase—the choice is arbitrary. In the analysis to be discussed below, the original intercalated clay was sealed in its aluminum cup, the top cover was carefully punctured with a sewing needle, and the cup was heated in a muffle furnace for several hours at about 300°C, a temperature above the deintercalation point, but below the onset of dehydroxylation. The deintercalated sample was then rerun, followed by a scan of the standard material (alumina or benzoic acid, depending on the temperature range), and the blank cup. Thus, the number of measurements increased by one over that needed for a simple C_p determination. Note that an event that spoils any of the four runs negates the entire experiment. Everything must work perfectly throughout what normally is a long and tedious series of measurements. For reasons discussed above, working at very low temperatures compounds the difficulties.

The method of determining C_p for intercalated molecules will be illustrated using analyses of Spruce Pine halloysite that had been intercalated with N-methyl formamide. The DSC settings are given in Table 8. The scheme for the computation follows that

Table 8. Differential scanning calorimetry (DSC) settings for heat capacity analysis described in Table 10.

Setting	Value for DSC	Value for computation	Name used in spread sheet
Heating rate	2.50°/min	0.04167°/s	HEAT
Range[2]	0.2 mcal/s	0.00870 mcal/s/cm	RANGE
Sample wt.	13.95 mg	13.95 mg	SAMPLE
Sample wt. (reweighed)	11.95 mg	11.95 mg	HEATED
Standard wt. (benzoic acid)	14.98 mg	14.98 mg	STANDARD

[1]Values for machine and values used in computations are shown along with symbolic names used in spread-sheet computations.
[2]Full scale is 23 cm.

outlined above, except that the final step involves the relation to give the heat capacity of the guest material.

Computations. In addition to the DSC settings, the following quantities were also calculated:

SCALE	((RANGE/23.)/HEAT)
MWSTAND (intercalated clay)	317.12
MWHEATED (deintercalated clay)	258.12
MWGUEST (guest molecules)	59.00

Table 9 shows the formulae used in the spread-sheet calculations.

Table 9. Formulae used for heat capacity computations using data in Table 10.[1]

Column	Formula
G	(STANDARD/SAMPLE)*((B10+$E10)/($D10+$E10))*$F10*MWSTAND
H	(STANDARD/HEATED)*((B10+$E10)/($D10+$E10))*$F10*MWHEATD
I	(G10/MSWTAND)-(HEATED/SAMPLE)*(H10/MMWHEATED)/(1-HEATED)/SAMPLE)*MWGUEST

[1]Symbolic names are defined in Table 8.

Table 10. Data obtained from heat capacity measurements from a Spruce Pine halloysite to illustrate the computations involved. [Data are presented as a spread sheet; cell addresses are designated by column (letter) and row (number). Formulae for computations that generated non-data columns are listed in Table 9.]

	A	B	C	D	E	F	G	H	I
1	111	1.42	-0.85	5.15	5.72	0.13323	29.8	20.0	10.3
2	121	2.28	0.03	6.03	5.53	0.14069	32.4	22.7	10.1
3	131	3.13	0.63	6.52	5.29	0.14806	36.0	24.8	11.6
4	141	3.88	1.22	7.06	5.20	0.15554	39.3	27.3	12.5
5	151	4.62	1.88	7.70	5.24	0.16300	42.3	30.0	12.8
6	161	4.90	2.20	7.89	5.40	0.17056	45.0	32.7	12.8
7	171	5.93	2.84	8.62	5.24	0.17829	49.0	34.8	14.7
8	181	6.58	3.48	9.33	5.10	0.18613	51.3	37.1	14.8
9	191	7.30	4.05	9.81	4.99	0.19408	54.9	39.7	15.8
10	201	8.02	4.50	10.49	5.00	0.20220	57.9	41.5	17.0
11	211	8.48	4.90	10.89	4.80	0.21047	60.7	43.6	17.8
12	221	9.10	5.21	11.48	4.98	0.21889	63.8	45.4	19.1
13	231	9.77	5.72	12.24	4.87	0.22747	66.3	47.1	19.9
14	241	10.26	6.20	12.75	4.75	0.23617	69.0	49.5	20.3
	T (K)	Sample Interc. cm	Sample Deint. cm	Standard cm	Blank cm	Standard C_p cal/g-K	Sample Interc. C_p cal/g-K	Sample Deint. C_p cal/g-K	Guest C_p cal/g-K

DISCUSSION

The foregoing sections have described the theory, practice, and implementation of several thermoanalytical techniques. The following two sections describe the results of two studies, each of which illustrates the types of problems that can be treated. One is a discussion of the desorption kinetics of N-methyl formamide intercalated into several different kaolinite samples. The second involves the measurement of the heat capacity of water intercalated between layers of kaolinite.

Deintercalation of N-methyl formamide

A common observation is that different kaolinite samples will intercalate a given material to different degrees (Sow, 1980; Tchoubar *et al.*, 1982) and at different rates. Olejnik *et al.* (1968) examined the rates of intercalation of kaolinite by dimethylsulfoxide, and Hach-Ali and Weiss (1969) determined the activation energy for the intercalation of N-methyl formamide. No comparable study apparently exists for the rates, kinetics law, or activation energy of deintercalation. This might be a useful way of studying kaolinite samples in general, because intercalation may be related to many of the important properties of kaolinite, such as degree of crystallinity (Tchoubar *et al.*, 1982).

As indicated above, examination of the kinetics of deintercalation is best achieved using a DSC to record the vaporization of the adsorbed phase. Any time a reaction occurs in which one or more of the products remains in the vicinity of the still-reacting material, the reaction rate may be influenced by the rate of product dispersion. If the product accumulates, the reaction will slow down and the kinetics will change. The presence of water vapor, for example, will retard the dehydroxylation of kaolinite (Holt *et al.*, 1962).

With the DSC, no simple way exists to ensure the rapid and complete removal of vapor during reaction. The best that can be done is to perform all experiments under as equal conditions as possible and to use the results in a comparative way, rather than in an absolute sense. The following sections deal with: (1) the selection of the correct rate law, and (2) the measurement of both the enthalpy of the desorption and the activation energy.

Samples. Several kaolin-group minerals were each intercalated to the maximum extent possible by treating them with N-methyl formamide liquid at 60°C, followed by air drying at ambient temperature for several days. All were at least 90% intercalated as indicated by X-ray powder diffraction analysis. For all samples, the DSC strip chart record was analyzed using all the rate laws discussed above; for the powder and Avrami-Erofeev models, several values of the exponent were tried—a total of eight determinations for each sample.

The results of Arrhenius plots for one of the aforementioned samples are presented in Figure 7. Inspection of these plots shows that (1) the fit to a straight line ranges from very good to mediocre, and (2) the activation energy determined from each plot also differs according to the rate law. In judging the goodness of fit, the points at the extremes of the temperature range should be ignored because they correspond to either the very onset of reaction or to the last moments of reaction, both of which are unlikely to be described by the same kinetics as the bulk of the reaction. Although some rate laws are obviously incorrect (e.g., the positive slope in Figure 7J), visual inspection of the Arrhenius plots fails to indicate a clear choice of rate law. The calculated activation energies range from almost 69 kJ/mole to about 8 kJ/mole. For these experiments, the weight loss at constant temperature was not measured (as described above); hence, that information was of no help in choosing the rate law.

The samples used and their identification codes are listed below:

Sample	Identification Code
Georgia, well-crystallized kaolinite	GNMR, GNMR-1, GN12780
Spruce Pine halloysite	SNMF2-1,SF12180A, SF12180, SF1980
Cornwall kaolinite	(RLO1415)RNNMR1, RNNMR
Dickite	DNMF3-1, DN12980, DNF3
Eureka halloysite	ENMR1

Deintercalation results. Comparison of the Arrhenius plots for all samples revealed that only the Avrami-Erofeev rate law with an exponent of 2 gave a good fit for all samples. This law was also identified by Adams (1978) in his study of an N-methyl formamide intercalate of kaolinite. Inasmuch as the computation of the kinetics automatically gives the enthalpy for the reaction, the two quantities (activation energy and enthalpy) are plotted for each sample in Figure 8. The scatter of points is large and unsystematic, but careful examination shows that the activation energy does indeed separate the samples into classes according to the parent clay. These groupings are indicated by the vertical lines in the figure, which delimit the halloysite samples (region A) from the Georgia kaolinite (region B) from the Cornwall kaolinite (region C) from the dickite (region D). For the halloysite samples, the range of enthalpy is large, however, this range is much less for the other groups. Why the halloysite samples reacted so differently from kaolinite and dickite when their N-methyl formamide intercalate decomposed on heating is unclear at this point, but may be related to the structural disorder common to all halloysite samples.

Heat capacity of water intercalated in kaolinite

The interaction of water with the surface of kaolin-group minerals has attracted much interest, focused chiefly on halloysite and its different hydration states. Halloysite is a less-than-ideal material to study, because it varies from sample to sample in terms of morphology, crystallinity, and degree of hydration. An alternative system was studied by Costanzo *et al.* (1980, 1984) consisting of well-crystallized kaolinite that had been intercalated with water. In this system, several hydrated states were recognized: two monohydrates, with d(001) = 8.4 and 8.6 Å, and two dihydrates, with d(001) = 10 Å. The monohydrates differed in that the 8.6-Å mate-

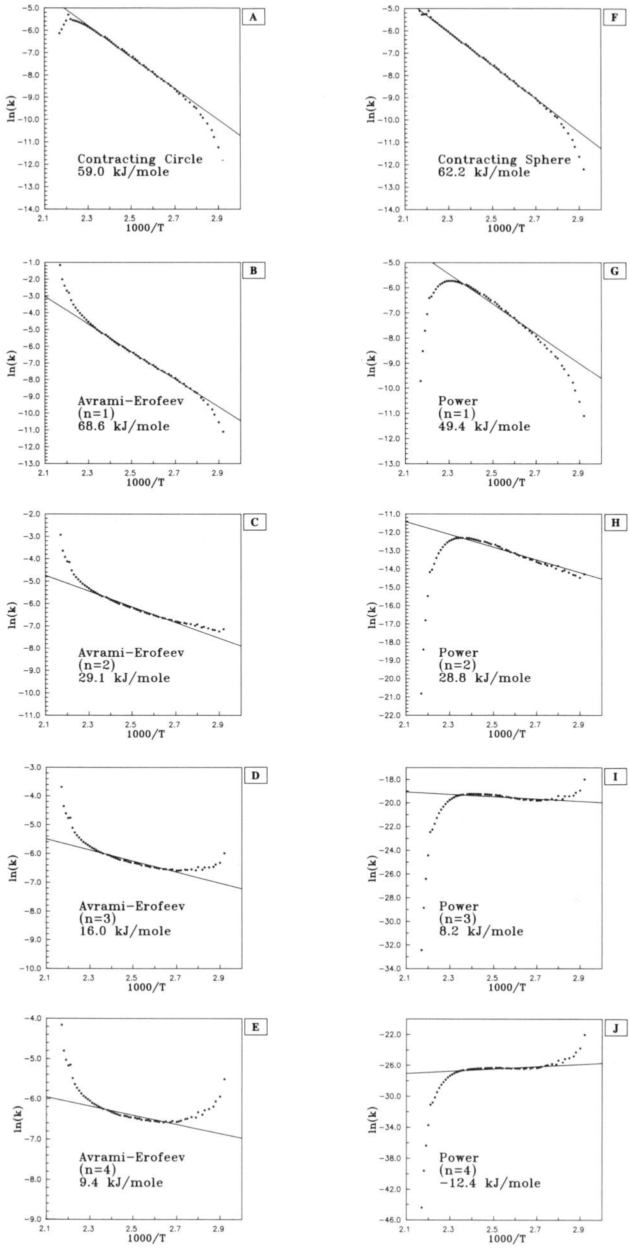

Figure 7. Arrhenius plots for the deintercalation of Georgia kaolinite-N-methyl formamide; (A) contracting circle law; (B) contracting sphere law; (C) Avrami-Erofeev law (exponent of 1); (D) power law (exponent of 1); (E) Avrami-Erofeev law (exponent of 2); (F) power law (exponent of 2); (G) Avrami-Erofeev law (exponent of 3); (H) power law (exponent of 3); (I) Avrami-Erofeev law (exponent of 4); (J) power law (exponent of 4).

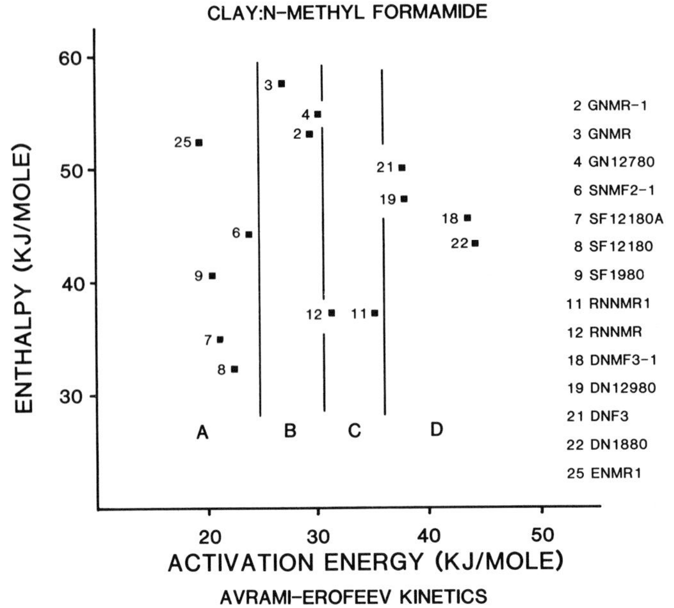

Figure 8. Plot of activation energy (x-axis; Avrami-Erofeev kinetics, exponent of 2) for deintercalation vs. enthalpy of the reaction (y-axis). Samples are kaolins intercalated with N-methyl formamide.

rial was disordered in terms of stacking,whereas the 8.4-Å hydrate was three-dimensionally ordered, giving several hkl reflections. The dihydrates differed in that one was similar to fully hydrated halloysite, whereas the other (termed QS-10) was more stable and could be kept fully hydrated under ambient conditions if a little care was exercised. These hydrates present, perhaps, a unique set of materials for the investigation of water in close proximity to a silicate surface.

The environment about the water molecule is very different in the 8.4-Å hydrate compared with the 10-Å hydrate. The former has isolated water molecules (termed "hole water"), which are keyed into the ditrigonal holes of the silicate tetrahedra. The 10-Å hydrate has two types of water: (1) hole-water molecules having the same geometric relation to the kaolinite surface as in the 8.4-Å hydrate, and (2) more loosely bound water (termed "associated water"), which is sandwiched between the hole-water molecules on one side and the hydroxyls of the adjacent kaolinite layer on the other. Given such a structure, the heat capacity of the two types of water should be very different.

The plots of C_p values obtained for the two hydrates (Figure 9; data from Costanzo *et al.*, 1982) show several interesting features. The experimental conditions were as described above, except that the heat sink was liquid nitrogen, which allowed measurements between 110 and 275 K. Values of C_p for both the 8.4- and 8.6-Å hydrates parallel that of pure water, with the more crystalline material having consistently smaller values of C_p that the disordered material. Such a difference is con-

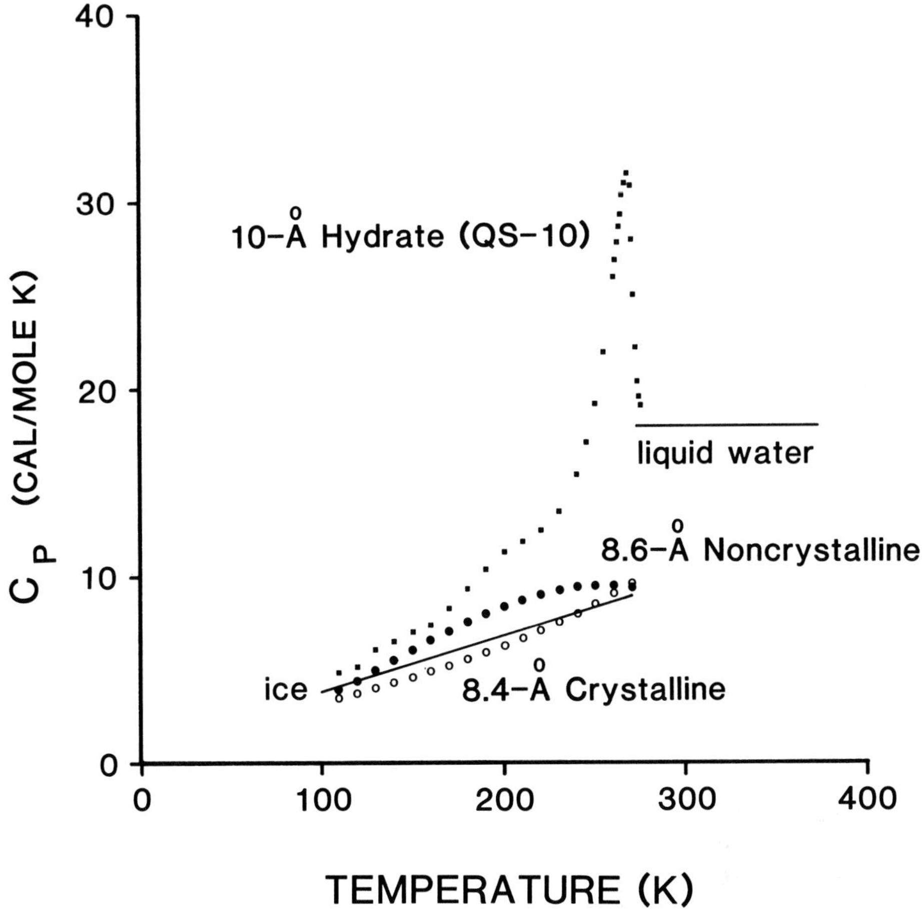

Figure 9. Heat capacity of water intercalated in kaolinite as a 10-Å hydrate (small, solid squares), as a disordered 8.6-Å hydrate (solid circles), and as an ordered 8.4-Å hydrate (open circles). Straight line segments are the heat capacity of ice and liquid water.

sistent with the slightly greater motional freedom of water in a restricted but disordered environment (8.6-Å hydrate) compared with water in a uniformly restricted setting (8.4-Å hydrate).

More structure is visible in the values of C_p for water in the QS-10 hydrate. At the lowest temperatures, the heat capacity is close to that of ice, but the intercalated water departs from the ice line at about 160 K—an event that is visible in the nuclear magnetic resonance T_1 signal as the onset of rotational motion (Costanzo *et al.*, 1982; Lipsicas *et al.*, 1985). C_p increases rapidly beginning at about 200 K and reaches a maximum at 271 K, falling abruptly thereafter to join the liquid water line. This anomaly has been interpreted as an order-disorder event involving the exchange of H_2O molecules among a set of possible sites in the interlayer region (Giese and Costanzo, 1986).

SUMMARY

The techniques described here for studying the kinetics of desorption and/or measuring heat capacity are tedious, but with the availability of small computers for collecting data and for controlling the DSC, this problem is somewhat reduced. A major difficulty remains, however, in that the experiments must be carefully prepared and the instrument must function properly and reproducibly during the course of what may be a long series of scans.

The techniques are underutilized at present; hence, the assessment of their ultimate utility is not possible at this time. So far, all the work seems to have been done using kaolin-group minerals. Similar studies of smectite minerals should produce interesting and potentially useful information as well.

REFERENCES CITED

Adams, J. M. (1978) Differential scanning calorimetric study of the kaolinite-N-methyl formamide intercalate: *Clays & Clay Minerals* **26,** 169-172.

Adams, J. M. and Watl, G. (1980) Thermal decomposition of kaolinite:dimethyl sulfoxide intercalate: *Clays & Clay Minerals* **28,** 130-134.

Costanzo, P. M., Clemency, C. V., and Giese, R..F. (1980) Low-temperature synthesis of a 10-Å hydrate of kaolinite using dimethylsulfoxide and ammonium fluoride: *Clays & Clay Minerals* **28,** 155-156.

Costanzo, P. M., Giese, R. F., and Lipsicas, M. (1984) Static and dynamic structure of water in hydrated kaolinites: Part I. The static structure: *Clays & Clay Minerals* **32,** 419-428.

Costanzo, P. M., Giese, R F., Lipsicas, M., and Straley, C. (1982) Synthesis of a quasi-stable kaolinite 10-Å and the heat capacity of the interlayer water: *Nature* **296,** 549-552.

Cruz, M. I., Letellier, M., and Fripiat, J. J. (1978) NMR study of adsorbed water. II. Molecular motions in the monolayer hydrate of halloysite: *J. Phys. Chem.* **69,** 2018-2027.

Giese, R. F. and Costanzo, P. M.(1986) Behavior of water on the surface of kaolin minerals: in *Geochemical Processes at Mineral Surfaces,* J. A. Davis and K. F. Hayes, eds., *ACS Symposium Series* **323,** American Chemical Society, Washington, D.C., 37-53.

Ginnings, D. C. and Furukawa, G. T. (1953) Heat capacity standards for the range 14 to 1200°K: *J. Amer. Chem. Soc.* **75,** 522-527.

Guarini, G. G. T. and Spinicci, R. (1972) DSC study of the kinetics of thermal dehydration of $BaCl_2 \cdot 2H_2O$ and $BaCl_2 \cdot H_2O$: *J. Therm. Anal.* **4,** 435-450.

Hach-Ali, P. F. and Weiss, A. (1969) Estudio de la reaccion de caolinita y N-metilformamida: *An. Soc. Quim. Argent.* **65,** 769-790.

Holt, J. B., Cutler, I. B., and Wadsworth, M. E. (1962) Rate of thermal dehydration of kaolinite in a vacuum: *J. Amer. Ceram. Soc.* **45,** 133-136.

Lipsicas, M., Straley, C., Costanzo, P. M., and Giese, R. F. (1985) Static and dynamic structure of water in hydrated kaolinites: Part II. The dynamic structure: *J. Coll. Interface Sci.* **107,** 221-230.

Lipsicas, M., Raythatha, R., Giese, R. F., and Costanzo, P. M. (1986) Molecular motions, surface interactions, and stacking disorder in kaolinite intercalates: *Clays & Clay Minerals* **34,** 635-644.

MacEwan, D. M. C. and Wilson, M. J.(1980) Interlayer and intercalation complexes of clay minerals: in *Crystal Structures of Clay Minerals and their X-ray Identification,* G. W. Brindley and G. Brown, eds., Mineralogical Society, London, 197-248.

Mackenzie, R. C. (1957) Thermal methods: in *The Differential Thermal Analysis of Clays,* R. C. Mackenzie, ed., Mineralogical Society, London, 1-22.

Mackenzie, R C (1981) Thermoanalytical methods in clay studies: in *Advanced Techniqes for Clay Mineral Analysis,* J. J. Fripiat, ed., Elsevier, New York, 5-29.

Olejnik, S., Aylmore, L. A. G., Posner, A. M., and Quirk, J. P. (1968) Infrared spectra of kaolin mineral-dimethyl sulfoxide complexes: *J. Phys. Chem.* **72,** 241-249.

Press, W. H., Flannery, B. P. Teukolsky, S. A., and Vetterling, W. T. (1986) *Numerical Recipes:* Cambridge University Press, New York, 818 pp.

Sow, C. (1980) Caracteristiques structurales des kaolinites des Charentes. Relation avec les proprietes spectroscopiques et l'aptitude a l'expansion: Theses 3[e] cycle, Universite d'Orleans, Orleans, France, 67 pp.

Sposito, G. and Prost, R. (1982) Structure of water adsorbed on smectites: *Chem. Rev.* **82,** 553-573.

Tchoubar, C., Plançon, A. Ben Brahim, H., Clinard, C., and Sow, C. (1982) Characteristiques structurales des kaolinites desordonees: *Bull. Mineral.* **105,** 477-491.

Theng, B. K. G. (1974) *The Chemistry of Clay-Organic Reactions:* Wiley, New York, 343 pp.

Thomas, J. M. and Clarke, T. A. (1986) The evaluation of kinetic parameter from thermoanalytical techniques. Dehydration of manganese formate dihydrate: *J. Chem. Soc. (A),* 457-460.

Wendlandt, W. W. (1986) *Thermal Analysis:* 3rd ed., Wiley, New York, 814 pp.

Young, D. A. (1966) *Decomposition of Solids:* Pergamon Press, New York, 209 pp.

HIGH-PRESSURE DIFFERENTIAL THERMAL ANALYSIS: APPLICATIONS TO CLAY MINERALS

A. F. Koster van Groos and Stephen Guggenheim

CONTENTS

Introduction	50
Experimental method	51
Apparatus	52
Sample preparation and characterization	53
Differential thermal analysis	54
Additional experimental considerations	58
Discussion	61
Structural aspects of phyllosilicates	61
Dehydration reactions in montmorillonite	64
Interpretation of HP-DTA data	64
Results	67
Physical model	73
Application to geologic systems	74
Dehydroxylation reactions (dioctahedral Al-rich phases)	75
Physical model for dehydroxylation in 2:1 layers	75
Interpretation of HP-DTA data	78
Results and discussion - kaolinite	78
Results and discussion - montmorillonite	86
Future of High-Pressure Differential Thermal Analysis	90
Acknowledgments	91
References Cited	91

HIGH-PRESSURE DIFFERENTIAL THERMAL ANALYSIS: APPLICATION TO CLAY MINERALS

A. F. KOSTER VAN GROOS AND STEPHEN GUGGENHEIM

Department of Geological Sciences, University of Illinois at Chicago
Chicago, Illinois 60680

INTRODUCTION

Thermal analysis of clay and other substances at atmospheric pressure has developed into a major analytical method, beginning with the pioneering studies of Le Chatelier (1887) on kaolinite, more than a century ago. Thermal analytical methods differ from chemical or structural methods in that they rely on a phenomenological approach by investigating the response of material with respect to a change in temperature. If a material is investigated at constant pressure, but at different temperatures, the results represent observations using one independent variable (i.e., temperature). If successive experiments are carried out at different pressures as well, a two-dimensional pressure-temperature grid can be created, from which important information may be derived.

The advantage of using pressure as an additional variable has not gone unnoticed, and different types of apparatus that allow high-pressure studies to be made have been developed (Wendlandt, 1986). The paucity of thermal analytical research at elevated pressures, however, suggests that these earlier methods are cumbersome. Recently, a relatively convenient high-pressure ($\geq$ 10 kbar) differential thermal analysis method was developed in the authors' laboratory (Koster van Groos, 1979). Because the apparatus is expensive to build and maintain, a simplified computer-controlled version is being developed for routine studies at moderate pressures ($<$2 kbar), which should facilitate wider usage of pressure in thermal analyses.

Rather than review the literature extensively or describe the technique in detail, the purpose of this chapter is to illustrate that high-pressure differential thermal analysis (HP-DTA) is an important tool in solving geologic problems. One such problem recognized as fundamental in mineralogy and petrology is the role of water in reactions involving clays and other phyllosilicate minerals. Recognizing how these materials act and interact as environmental conditions change requires a clear understanding of these material from both the thermodynamic and atomistic points of view. Therefore, instead of presenting HP-DTA data only, this chapter integrates structural information and derived kinetic data to relate results to atomistic models of decomposition. This type of presentation is possible because run conditions and reaction mechanisms may be assessed by observing trends in the DTA peak characteristics as pressure varies. Such an approach is not possible using traditional thermoanalytical methodologies at a single pressure.

Because this chapter introduces a technique and presents an integrated approach to understanding the thermal stability of phyllosilicates, it is divided into two general sections: methodology and discussion. The latter section covers a general discussion of phyllosilicate structures, the presentation and interpretation of data, and structural and thermodynamic relationships. Because the thermal stability of phyllosilicates involves many different reactions, the experimental data are presented intermixed with discussion. The potential of HP-DTA as an analytical and research tool will become evident in the course of this discussion.

EXPERIMENTAL METHOD

In HP-DTA studies a variety of equipment has been used, such as externally heated pressure vessels for experiments to 4 kbar (Harker, 1964), internally heated pressure vessels for studies to 10 kbar (Yoder, 1950; Eggler and Rosenhauer, 1978), and piston-cylinder apparatus for pressures from 10 to 40 kbar (e.g., Cohen and Klement, 1967). DTA studies of clay minerals using piston-cylinder apparatus (e.g., Khitarov and Pugin, 1966), however, are uncommon and are not discussed further in this chapter.

In Harker's (1964) method, two sealed Pt capsules containing reference material and sample, respectively, are connected by a short $Pt_{90}Rh_{10}$ wire. By welding a Pt wire onto each capsule and attaching a $Pt_{90}Rh_{10}$ wire to the connecting $Pt_{90}Rh_{10}$ wire, a DTA cell is formed. The temperature is given by the Pt-$Pt_{90}Rh_{10}$ thermocouple circuit of the reference capsule, and the differential signal is provided by the two Pt wires. Advantages of this method are the intimate contact of the thermocouple junction with the reference and sample and the possibility of obtaining data from closed systems. Although quality data have been obtained using this technique, several serious problems are apparent. Large thermal gradients may develop within the sample and between the sample and the thermocouple junction because of the relatively low thermal conductivity of platinum. Spurious and large fluctuations in the baseline may result from convection in the Ar gas pressure-medium, especially in the absence of an enclosing cell. Consequently, the sensitivity of the system is compromised by large and unpredictable temperature variations between sample and reference. In addition, assembly of the cell is complex and the geometry is difficult to reproduce between runs, which reduces the precision of the data.

Significant improvements were made by Eggler and Rosenhauer (1978), who used a large Pt block containing a sleeve, spot-welded onto a thermocouple, in which a Pt sample capsule could be placed. The reference thermocouple was placed directly in an aluminum space filler in the furnace assembly. With this system they obtained results at temperatures to 1480°C and pressures to 8 kbar. The weight of the samples was not specified, but the size of the capsules, 4 x 20 mm, suggests a sample weight of 100-200 mg. As no DTA patterns were shown in their paper, the quality of the data is difficult to assess; however, the published results appear to be very good.

A HP-DTA apparatus was developed (Koster van Groos, 1979) that combines ease of use with high-quality results. It currently allows studies to 950°C and 10 kbar, but with simple modifications the temperature range can be extended to 1400°C. At present, several hundred runs have been made without serious problems.

Apparatus

The main features of the apparatus were described by Koster van Groos (1979). As shown in Figure 1, the DTA cell consists of a Cu block containing positions for three capsules. The capsules are fabricated uniformly of Au using extrusion techniques, with a length of 10 mm, a diameter of 3.2 mm, a wall thickness of 0.05 mm, and a weight of about 120 mg. A reentry well about 1 mm deep is present in each capsule to accommodate a thermocouple (usually made of Pt-$Pt_{90}Rh_{10}$), which is spring-loaded to ensure thermal contact. Sample material is placed in two capsules, and the third contains either TiO_2 or Al_2O_3 reference material. The Cu block and thermocouple assembly are placed in an internally heated pressure vessel similar to the one described by Holloway (1971).

Sample temperature is obtained from the reference temperature and corrected using the differential temperature. The correction can be as much as 3°C, but is usually less than 1°C. Because of this correction, the nature of the reference is unimportant as long as it is inert in the temperature range of interest. DTA signals are recorded using the 20-μV or 50-μV range of a strip-chart recorder that can detect temperature deviations as small as 0.02°C. Pressures to 5 kbar are measured to within 0.5%, using low-, medium-, and high-pressure Bourdon-type calibrated Heise gauges. At pressures >5 kbar, a Harwood manganin cell calibrated against the Heise gauges is used. In most experiments the heating rate is held at 20°C/min using a programmable controller. The system is regularly calibrated at different pressures using the low-quartz-to-high-quartz inversion (Yoder, 1950; Koster van Groos and ter Heege, 1973). Corrected temperatures are believed to be accurate to $\pm$ 1°C.

An important feature of the method described here is that the geometry of the capsules and their position in the cell are reproducible, allowing comparison of DTA signals. Furthermore, the capsules may be either open or welded shut, thus making studies in closed systems possible. The purpose of the reentry well is not as much for positioning the thermocouple with respect to the capsule as for placing the thermocouple near the center of mass of the sample. Using this sample geometry, the onset and termination of the peak representing the low-high transition of 4 mg of quartz, well-mixed with a 20-mg sample, extends over a 3°C temperature interval with a sharply defined peak near the center. No visible effects of signal broadening have been detected; thus, thermal gradients within the sample are probably small

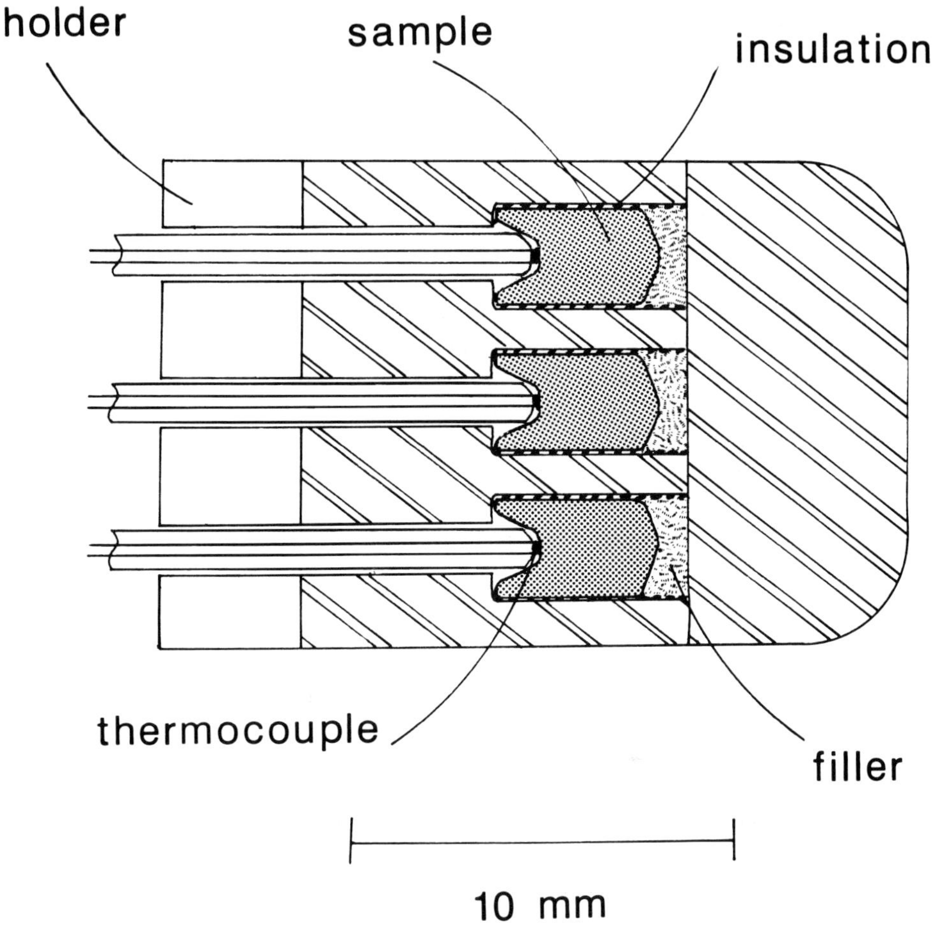

Figure 1. Cu differential thermal analysis cell containing two positions for samples and one position for a reference.

(possibly <1°C at 573°C). The reproducibility of the thermograms is excellent, to within 0.5°C.

Sample preparation and characterization

For studies of phyllosilicates or clay materials, sample preparation for HP-DTA is usually minimal. If the starting material is coarse grained, size reduction by abrading the material with an Fe file is advisable to avoid damage to the crystalline structure, as is often the result of dry grinding. Possible Fe contamination is easily removed with a magnet. All starting material should be sieved to pass 200 mesh;

expandable clays should be stored over a saturated solution of $Mg(NO_3)_2{\cdot}H_2O$ (55% relative humidity); other samples are stored in a desiccator.

Unless otherwise specified, the starting materials used in the studies described below were: (1) kaolinite from the Coss Hedger pit, near Sandersville, Washington County, Georgia, and (2) Na-rich montmorillonite from the Newcastle Formation, Crook County, Wyoming (Source Clays Repository of The Clay Minerals Society, samples KGa-1 and SWy-1, respectively; see van Olphen and Fripiat, 1979).

Sample preparation methods depend on the nature of a study. One general concern, however, is the effect of grain size on the DTA signal. To determine such effects in montmorillonite, a series of DTA experiments were made with sample SWy-1 at 1 bar in the HP-DTA apparatus (Koster van Groos and Guggenheim, 1984). Figure 2 shows DTA patterns of the SWy-1 sample and eight different hydraulic size fractions, ranging from >10 to <0.1 μm. Most contaminants were removed in sizing the sample, and the finest fraction only contained some colloidal SiO_2 in addition to the montmorillonite. The results were unaffected by particle size, although the DTA peak of the finest fraction (<1 μm) was somewhat sharper than for the other fractions. Therefore, in experiments on the effect of different interlayer cations on dehydration, only the finest fraction was used. Note, however, that particle size and hydraulic fraction could not necessarily be equated. For example, scanning electron microscopy (SEM) showed that the coarse fractions of sample SWy-1 are composed of aggregates of smaller crystals and that crystal size does not vary significantly between the different fractions (Koster van Groos, unpublished).

To determine the effect of interlayer cation on water-loss or hydroxyl-loss reactions, sample SWy-1 was homoionically exchanged with K, Ca, or Mg by four washings of the clay with 1 N chloride solutions at ambient conditions, followed by six washings with distilled water and drying in air. These exchanged montmorillonite materials will be referred to as K-SWy-1, Ca-SWy- 1, and Mg-SWy-1.

In all HP-DTA studies, characterization of starting materials by X-ray powder diffraction and chemical analysis is advisable to determine the mineralogical and chemical compositions. In the present studies, fused samples (20 mg of clay fused on a Pt wire loop at 1200°C for 1 min) were analyzed by electron microprobe to determine chemical composition. Results were satisfactory, although some Na was lost during fusion (Koster van Groos and Guggenheim, 1984).

Differential thermal analysis

The interpretation of a DTA pattern is still a subject of discussion in the thermal analysis literature. The temperature of the extrapolated onset of a peak, the peak temperature, the extrapolated peak temperature using the tangent of the main slope of the peak, the temperature of the intersection of the extrapolated limbs of a peak, or other segments of the peak all have been used to determine the reaction

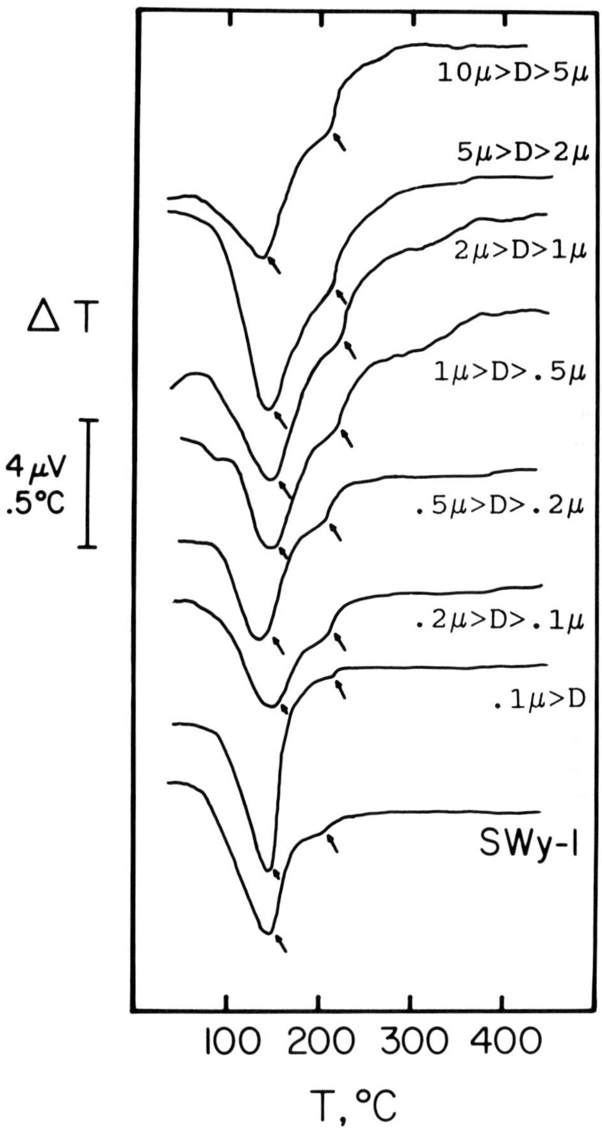

Figure 2. Differential thermal analysis patterns for sample SWy-1 and for different hydraulic size fractions (D) of sample SWy-1, ranging from >10 to <0.1 μm (Koster van Groos and Guggenheim, 1984). Arrows indicate temperature at which dehydration reactions were almost complete.

temperature, the enthalpy of reaction, and information on the kinetics of the reaction. The consensus is, however, that the reaction rate is greatest at the peak temperature.

To assess the many variables that determine the relation between a reaction and its DTA peak pattern without considering the nature of the reaction is difficult. For example, in a single inversion reaction, such as the low-quartz-to-high-quartz inversion, the activity of the phases is either unaffected or only slightly affected during the reaction. During a dehydration, dehydroxylation, or decarbonation reaction, however, the fugacity of the evolving gas phase may increase during the reaction, which, in general, causes the reaction temperature to increase, resulting in a broadening of the peak. Therefore, possible changes in the conditions during thermal analysis and the effect of these changes on the observations must be considered.

For the interpretation of DTA and other thermal analytical data, whether a reaction represents stable or metastable equilibrium is not important. Reversibility of such a reaction is insufficient evidence to determine the nature of the equilibria because a metastable reaction may be readily reversible, e.g., the partial dehydroxylation of muscovite (Heller *et al.*, 1962). In general, evidence from long-duration experiments is necessary to determine the nature of the reaction.

If more than one reaction occurs at about the same temperature, the peaks in a thermogram will interfere with each other (see Figure 3). Curve 1 in Figure 3A shows a simplified dehydration peak of smectite. The extrapolated onset (a), the peak (b), and the point at which the reaction is completed (c) define the signal. Curve 2 shows the results when a second reaction occurs simultaneously, such as the boiling reaction of water, hereafter referred to as $(L,V)_{H_2O}$. The dashed lines show the individual peaks. To determine the onset of the smectite dehydration peak on curve 2, the additional peak must be subtracted. Although deconvolution methods can be used for known or ideal peaks shapes, in real systems the size and shape of additional peaks are difficult to assess and the determination of the onset of the second reaction is virtually impossible. Fortunately, a rapid change in the reaction rate at the peak is normal and results in a well-defined inflection point in the peak-limb. Its temperature represents the peak temperature of the second reaction.

In some materials the reaction is more complex. For example, in montmorillonite two dehydration peaks are often observed (Figure 4). This is illustrated ideally in Figure 3B, curve 1. With water present, the reaction $(L,V)_{H_2O}$ is encountered at (d) in curve 2, and the two dehydration reactions are indicated as shoulder peaks at the inflection points (c) and (e). This was illustrated in the study of sample SWy-1 (Koster van Groos and Guggenheim, 1984), in which two dehydration reactions occurred (Figure 2). In the presence of excess water, the $(L,V)_{H_2O}$ reaction is superposed on the two dehydration reactions of montmorillonite (Figure 5). It occurs at the lowest temperature and is represented by the largest peak. The peak temperatures of the two dehydration reactions are shown by breaks in the slope toward the baseline, similar to Figure 3B. A series of experiments at pressures to 50 bars con-

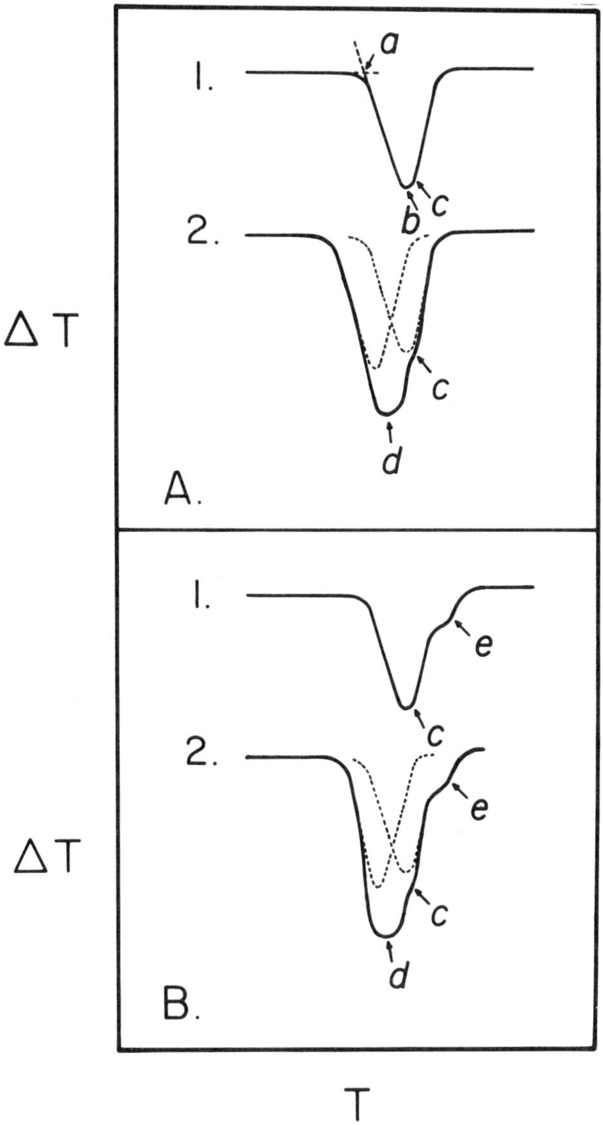

Figure 3. (A) Curve 1 shows idealized differential thermal analysis pattern of smecite. Point (a) represents onset of reaction; point (b), peak temperature; point (c), temperature at which reaction is almost complete and curve starts its sharp return to the baseline. Curve 2 shows effect of the presence of water. Point (d) represents the combined peak. Point (c) is a shoulder in the curve and is at the same temperature as point (c) in curve 1. (B) Idealized patterns as in (A), but the smectite has two dehydration reactions.

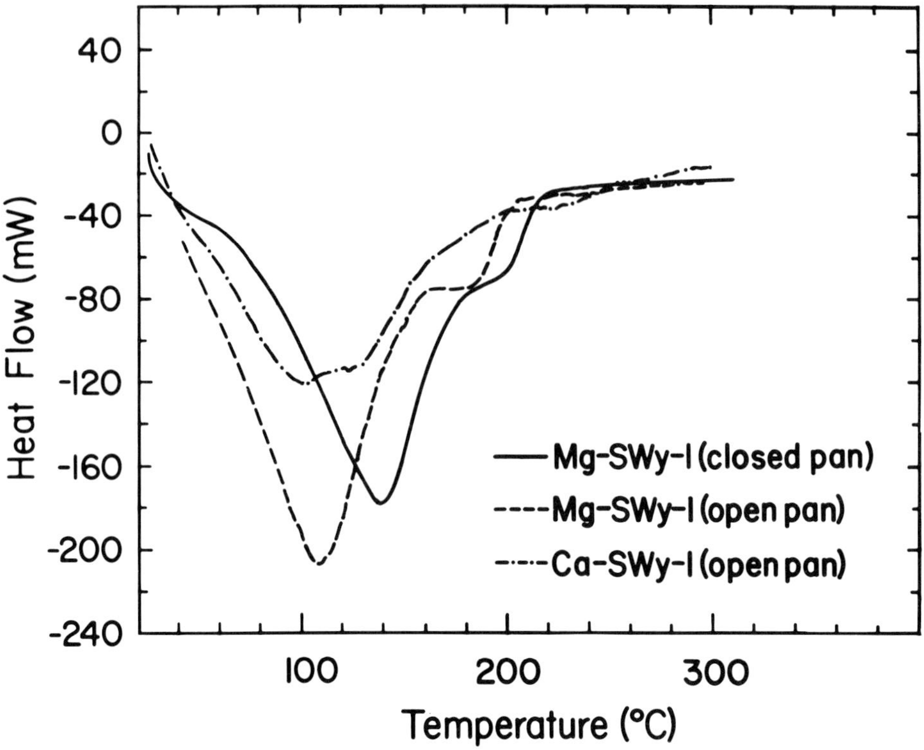

Figure 4. Differential scanning calorimetry patterns for samples Ca-SWy-1 and Mg-SWy-1 in N atmosphere, using a Dupont 1090 analyzer. Note two peaks for sample Ca-SWy-1 between 100° and 140°C. Analyst: P. van Krieke, University of Utrecht (data from Koster van Groos and Guggenheim, 1987a).

firmed (Koster van Groos and Guggenheim, 1984) that the temperature of these breaks in slope was reproducible and was the same as the peak temperature in experiments in which only smectite was present. On this basis, one can be confident that these temperatures do, in fact, represent actual reaction temperatures.

Additional experimental considerations

In DTA, the sample is commonly heated in air at 1 atm, and any vapor produced by the sample usually escapes. Some DTA methods facilitate this vapor release by directing a flow of an inert gas, such as N_2, over the sample. As will be seen, significant differences in dehydration temperatures of smectites may result from small differences in the experimental procedure because of the effect of varying H_2O fugacity, γ_{H_2O}. For example, heating larger and more compacted smectite samples may result in an increase of γ_{H_2O} within the sample during a dehydration reaction, which may raise the observed dehydration temperatures by several tens of degrees Celsius. Therefore, extreme care is necessary to ensure constant conditions and to understand the conditions present in the sample.

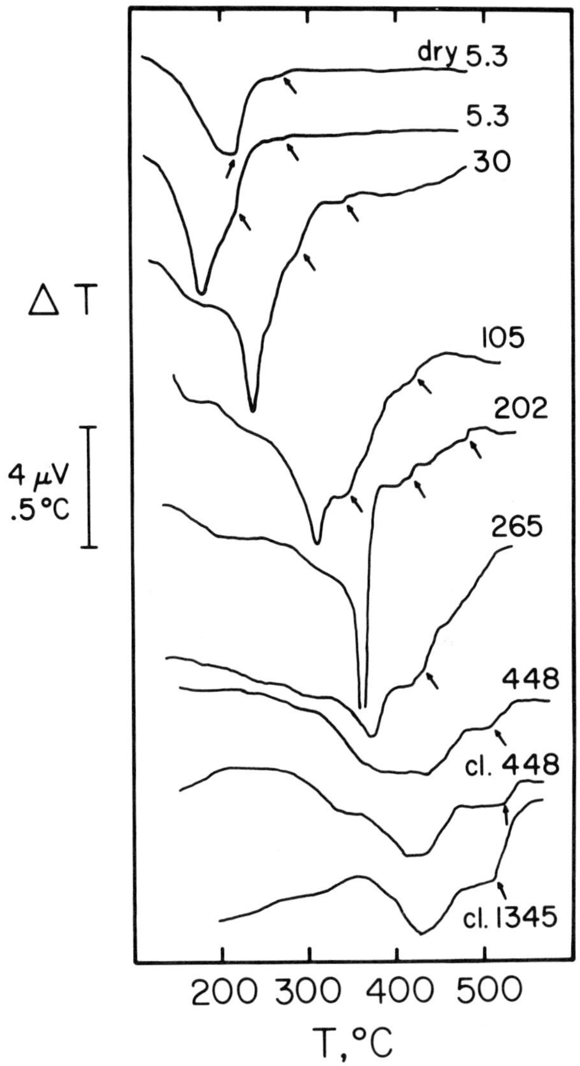

Figure 5. High-pressure differential thermal analysis patterns of sample SWy-1 using open and closed capsules. Arrows indicate temperatures at which dehydration is almost complete (cf. points (c), Figure 2). Unlabeled low-temperature peaks represent the reaction $(L,V)_{H_2O}$ (Koster van Groos and Guggenheim, 1984).

As currently designed, the HP-DTA system allows thermal analysis in open or closed systems at pressures to 10 kbar. In analyses using open capsules the composition of the vapor phase must be considered when investigating dehydration or dehydroxylation reactions because P_{H_2O} may not equal P_{total}. Koster van Groos and Guggenheim (1984) attempted to evaluate and control the composition of the vapor surrounding the sample. The molar volume of water vapor at low pressures and elevated temperatures is high. Consequently, sealed capsules tend to burst at these pressures. The volume of H_2O vapor generated during the early stages of dehydration of smectite in low-pressure experiments, however, appears sufficient to displace the Ar pressure medium in open capsules. Thus, the composition of the gas phase close to the sample approaches pure H_2O. To limit mixing of the evolving water vapor with the Ar pressure medium, which would reduce γ_{H_2O} near the sample, the space in the capsule above the samples was tightly filled with an inert fiber, such as quartz glass wool. In these experiments, conditions of $P_{H_2O} = P_{total}$ can be maintained at temperatures to about 300°C and pressures to about 50 bars.

In analyses at pressures between 50 and 200 bars, the introduction of 5-30 wt. % water at the bottom of the capsule before adding the sample was necessary to maintain $P_{H_2O} = P_{total}$. If insufficient water was added or if water was lost prior to or during a run, lower and non-reproducible dehydration temperatures were observed. In runs having sufficient water present, the strong endothermic peak of $(L,V)_{H_2O}$ interfere with the smectite peaks, as discussed above; however, the P-T conditions of $(L,V)_{H_2O}$ are well known (Keenan *et al.*, 1978), and the related peak was easily identified.

Closed capsules can be used successfully if the pressure within the capsule does not exceed the external pressure. In such experiments, typically 5-10 wt. % water was added to the starting material to ensure that $P_{H_2O} = P_{total}$. Above the critical pressure of water the liquid vapor transition occurs over a temperature range, and, thus, the resultant peak was observed over an increasingly wide temperature range with a further increase in pressure. The interpretation of the dehydration pattern of the runs at these conditions was occasionally problematic because of difficulties in the identification of dehydration peaks.

In dry runs, the density of the sample was controlled by compressing it with a constant force. In open-capsule runs with water present, water was pipeted into the capsules before the sample was introduced. In these runs the sample was added carefully, otherwise a soft clay-water mixture flowed out of the capsule during the compression step. Water had to be added to the capsule first to ensure that during heating the water vapor from the evaporating water enclosed the sample. In most runs about 20-30 mg of sample was used; the amounts of both water and sample were determined by weighing. Sealed capsules were checked for leaks before each experiment. If the capsule lost weight during the run, the results were not used.

DISCUSSION

Structural aspects of phyllosilicates

To understand the results of HP-DTA thoroughly, relevant aspects of the material investigated must be considered. The structures of phyllosilicate minerals (Figure 6) are dominated by planes of octahedrally coordinated and tetrahedrally coordinated cations. The octahedral cation and nearest neighboring anions comprise the octahedral sheet due to a two-dimensional linkage through the sharing of polyhedral edges. A hexagonal-based tetrahedral sheet is formed by the linkage of TO_4 groups (where T = Si, Al, or Fe), in which individual tetrahedra share three corners (basal oxygens) with tetrahedral neighbors. The hexagonal character of the sheet results from the two-dimensional linkage of tetrahedra, which form sheets composed of six-fold rings. The fourth tetrahedral corner (apical oxygen) is not involved in sharing corners with other tetrahedra, but is involved in the joining of the tetrahedral and octahedral sheets. The apical oxygen forms part of a common plane of junction between tetrahedral and octahedral sheets and belongs to both the octahedral cation coordination and the tetrahedral cation coordination. Different assemblages may form depending on the number of component sheets that link together. If one tetrahedral sheet is linked to an octahedral sheet, a 1:1 layer results, whereas if two tetrahedral sheets oppose one another with an octahedral sheet between, a 2:1 layer is created.

Although two apical oxygens per attached tetrahedral sheet form the coordination about each octahedral cation, additional anions, typically OH or less commonly F , are needed to complete six-fold coordination. For example, for 2:1 phyllosilicates, two additional OH groups are required. The OH groups are located on opposing sides of the octahedral cation with each at nearly the same level along the *c* axis as the apical oxygens and in the center of the six-fold ring. Consequently, the two OH groups are energetically very similar. In contrast, 1:1 phyllosilicates have only one attached tetrahedral sheet, and such a sheet can supply only one OH group per octahedral cation in an analogous manner. An additional plane of OH groups is therefore necessary to complete the 1:1 layer. Thus, in 1:1 phyllosilicates, two very different types of OH groups exist: those located within six-fold rings and others which reside on an internal surface parallel to (001). The latter form long (about 2.9 Å) hydrogen bonds across the interlayer to the basal oxygen of the silicate sheet of the next 1:1 layer. The location of the OH group in the structure and, therefore, its bonding energies determine its response to temperature and pressure changes.

If the 1:1 or 2:1 layers are electrostatically charged, adjacent layers may be separated by additional cations, hydrated cations, metal-hydroxyl octahedral sheets, organic compounds, or various other interlayer materials to offset the residual charge on the layer. In addition to the type of layer, phyllosilicates are often classified on the basis of this interlayer material. For example, 2:1 phyllosilicates containing interlayer material consisting of complete metal-hydroxyl octahedral sheets have what is

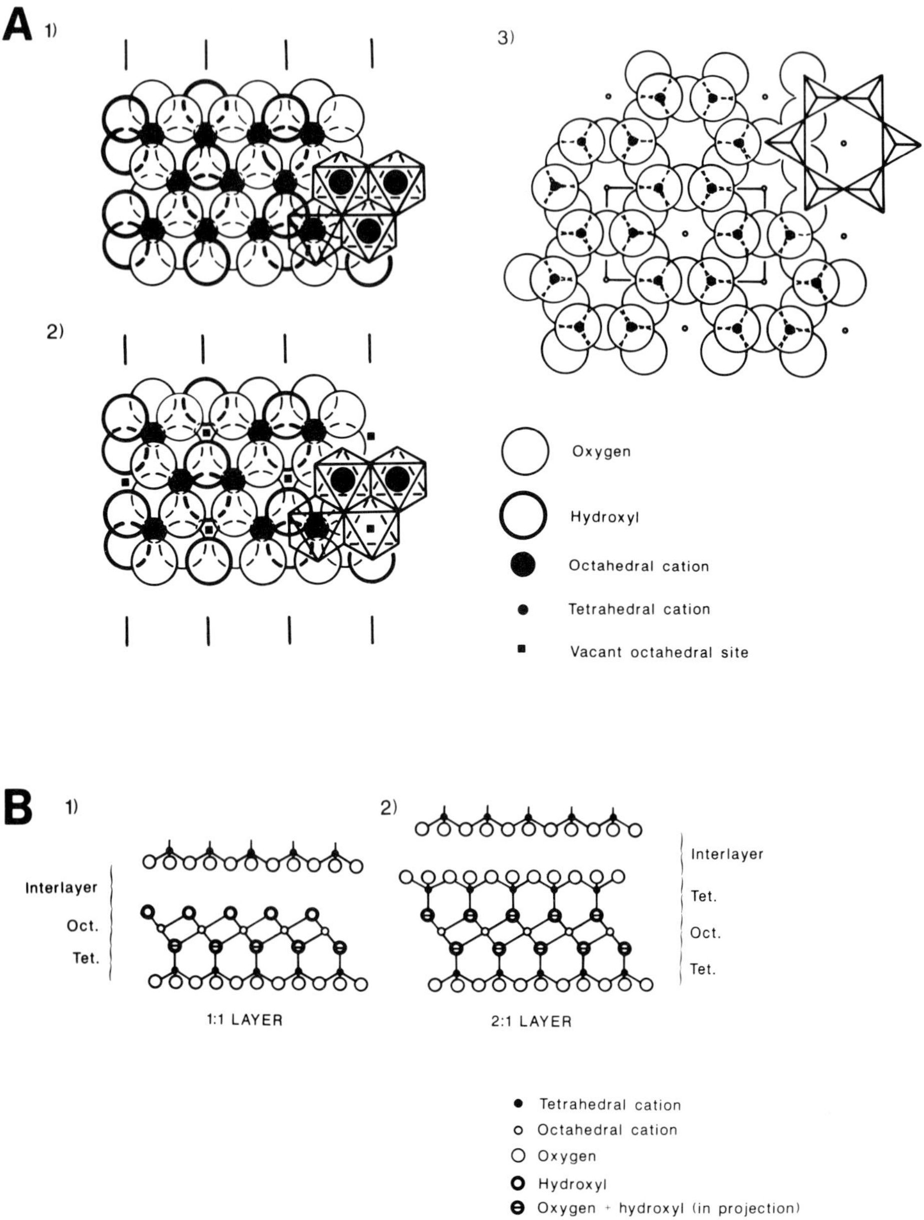

Figure 6. Structure of phyllosilicates (after Bailey, 1984).

called a chlorite structure. In contrast, if hydrated cations comprise the interlayer, a smectite structure results. Structures intermediate between chlorites and smectites, i.e., those containing cations hydrated with two H_2O planes forming an incomplete octahedral sheet configuration, are known as vermiculites (see Table 1).

Interlayer material is important in understanding thermal stability for two reasons. (1) Obviously, if H_2O molecules or OH groups are located in the interlayer region, the stability of these molecules will be different from those within the 1:1 or 2:1 layer. (2) Interlayer material may interact with the OH groups located within a 2:1 layer, as takes place with the interlayer cation in a mica, which resides partially within the silicate ring and is close enough to the OH groups to influence thermal behavior. In some materials, interlayer material may migrate from the interlayer region into the 2:1 layer, thereby affecting the thermal stability of the OH groups within a 2:1 layer (e.g., see Figure 6).

Commonly, H_2O is sorbed on the surface of fine particles because of the finite charge associated with broken bonds on crystal surfaces. Such surface effects are considerable for clay minerals because, in addition to the surfaces bounding the individual particle, some clay minerals have "internal" surfaces (an atomic plane at the interlayer as noted above), which are readily accessible to H_2O molecules residing in the interlayer. Usually, H_2O loss from either surface can be distinguished on the

Table 1. Classification of 1:1 and 2:1 phyllosilicates (after Bailey, 1984).

Layer type	Group[1]	Sub-group
1:1	Serpentine-kaolin	Serpentine
	(x ≈ 0)	Kaolin
2:1	Talc-pyrophyllite	Talc
	(x ≈ 0)	Pyrophyllite
	Smectite	Saponite
	(x ≈ 0.2-0.6)	Montmorillonite
	Vermiculite	Trioctahedral vermicuite
	(x ≈ 0.6-0.9)	Dioctahedral vermiculite
	Mica	Trioctahedral mica
	(x ≈ 1.0)	Dioctahedral mica
	Brittle mica	Trioctahedral brittle mica
	(x ≈ 2.0)	Dioctahedral brittle mica
	Chlorite	Dioctahedral chlorite
	(x is variable)	Di,trioctahedral chlorite

[1]x = charge per formula $O_{10}(OH)_2$ unit.

basis of the response to thermal effects; H_2O is lost from external planes at lower temperatures (< 100°C) than from internal planes.

Therefore, H_2O occurs in phyllosilicate either as OH groups (referred to below as strongly bonded H_2O) or as H_2O molecules. The former may be present at the junction between the tetrahedral and octahedral sheets in either 1:1 or 2:1 layers, they may be associated with the octahedra to complete six-fold coordination at an "internal" surface, or they may be associated with either a partial or complete octahedral interlayer. In contrast, H_2O molecules are located in the interlayer (hydrated cations) or on external surfaces and tend to be poorly bonded to the structure. Thermal analytical techniques offer a method of distinguishing between the various H_2O or OH constituents on the basis of difference in site energy as determined by their response to thermal effects. H_2O loss involving poorly bonded interlayer water is discussed below as a "dehydration" reaction, whereas loss of structural water (OH) is termed a "dehydroxylation" reaction. H_2O loss from external planes is not considered further.

Dehydration reactions in montmorillonite

In a series of studies on homoionically exchanged montmorillonite, Koster van Groos and Guggenheim (1984, 1986, 1987a) determined the effect of temperature and pressure on the dehydration reactions of sample SWy-1 before and after cation exchange with K, Ca, and Mg. HP-DTA is especially well suited for such studies because the dehydration reaction is reversible and non-quenchable, thereby making long-duration equilibrium-type studies difficult.

An increase in the fugacity of water results in a large increase in the temperature of the dehydration reaction. This increase in temperature is especially apparent at low pressures, as can be seen in Figure 4. A differential scanning calorimetry (DSC) analysis of the interlayer H_2O loss in sample Mg-SWy-1 in an anhydrous N_2 atmosphere was made using two identical samples. The only difference between the two runs was that in one the sample-holder lid was left ajar. In the closed pan, evolved water vapor resulted in a slightly higher P_{H_2O} around the sample, and peak temperatures were about 15°C greater than the other. Therefore, even small differences in P_{H_2O} pressure can produce significant differences in observed temperatures.

Interpretation of HP-DTA data. Figures 5, 7, and 8 show representative HP-DTA patterns for sample SWy-1 and Ca-, Mg-, and K-exchanged products of this material. In experiments at elevated pressures, the capsules were usually loaded with extra water to maintain conditions at $P_{total} = P_{H_2O}$ (*vide supra*). In these runs, the peak representing $(L,V)_{H_2O}$ dominated and was superposed (see Figure 3) over the peaks resulting from the dehydration of the montmorillonite. For open-capsule runs, the water peak was usually asymmetric due to evaporation during heating. When $(L,V)_{H_2O}$ was reached, the remaining water boiled to produce a sharp peak (e.g., 34-bar run Figure 7). Thus, the higher temperaure side of the peak was well defined.

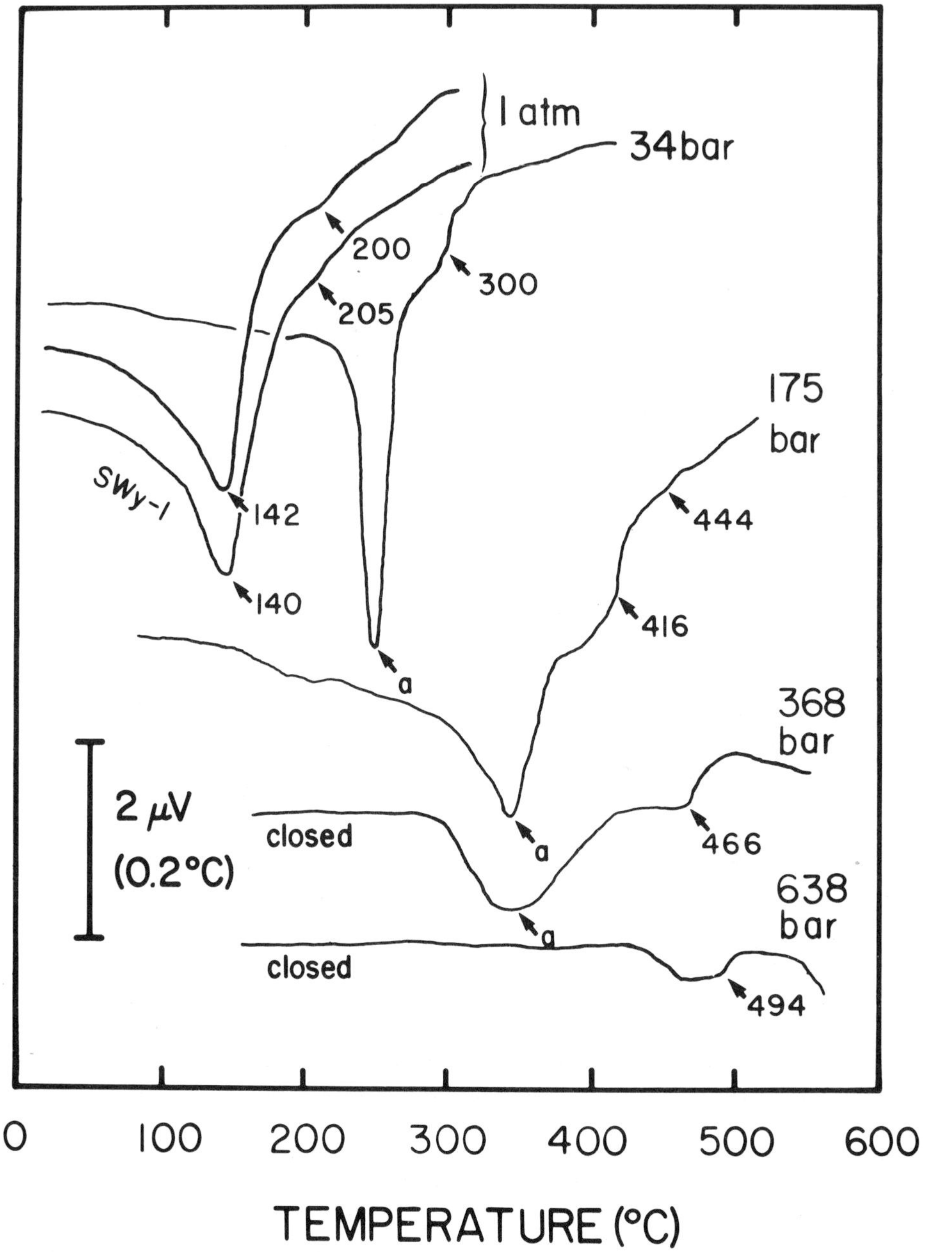

Figure 7. High-pressure differential thermal analysis patterns for dehydration reactions of sample K-SWy-1, using open and closed capsules. For comparison, a pattern of sample SWy-1 is included. Arrows indicate temperature at which each dehydration reaction is almost complete (cf. points "c" in Figure 2). Peaks labeled "a" represent the reaction $(L,V)_{H_2O}$ (Koster van Groos and Guggenheim, 1986).

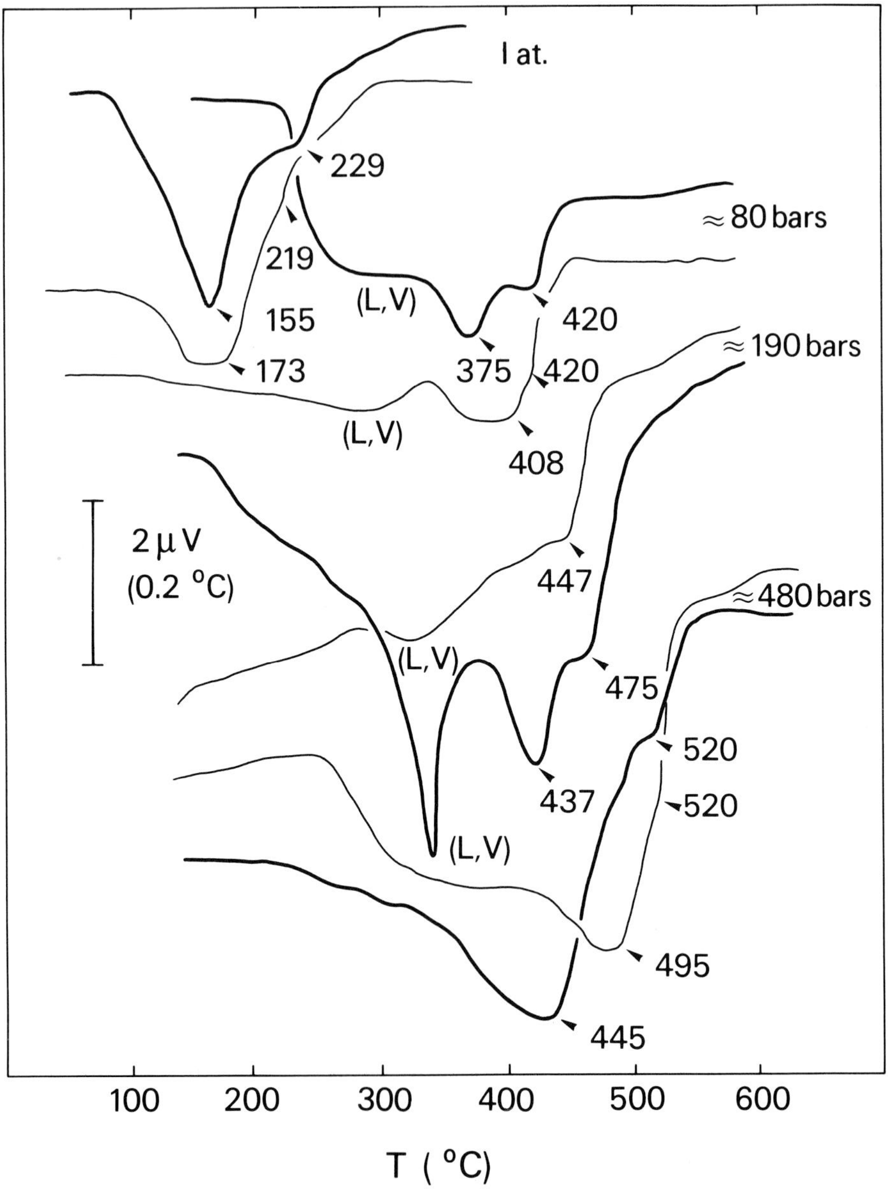

Figure 8. High-pressure differential thermal analysis pattern for dehydration reactions of samples Ca-SWy-1 (light lines) and Mg-SWy-1 heavy lines), using open and closed capsules. Arrows indicate temperature at which each dehydration reaction is almost complete (cf. points "c" in Figure 2). (Koster van Groos and Guggenheim, 1987a).

The shape of the $(L,V)_{H_2O}$ peaks for the closed-capsule runs was similar (e.g., 175 bars), unless the critical pressure of water was exceeded, in which case the peak was broader (e.g., runs at 368 and 638 bars).

In these patterns the dehydration reactions were identified with little ambiguity. In most patterns two peaks were present (e.g., Figure 8 at 190 bars), in addition to the $(L,V)_{H_2O}$ peak, although for some runs, only one peak was observed (e.g., at 80 bars). Note the significant differences between patterns from samples having different interlayer composition. For example, in runs with sample K-SWy-1 the peak for the second dehydration was much smaller than for the other samples.

Koster van Groos and Guggenheim (1984) suggested that the break in slope on the high-temperature side of the $(L,V)_{H_2O}$ curve or a small shoulder peak may be used as an indication that the dehydration reaction is essentially complete (see Figure 3). For runs in which the second dehydration peak is small, as for sample K-SWy-1, verification that the break in slope was consistent in different runs is important. Thus, all analyses in which $P_{H_2O} = P_{total}$ is assumed, i.e., dry samples at low pressures and analyses of wet samples at intermediate pressures and in sealed capsules at high pressures, must yield reproducible and consistent results.

Results. The peaks in Figures 5 and 6 are denoted by arrows. Usually, two peaks were noted: a peak at the higher temperature that is noticeably weaker than a peak at the lower temperature. Previous DTA and infrared (IR) evidence (e.g., Grim, 1968, p. 288; Farmer, 1971; Low, 1979) has suggested that interlayer water is bonded unequally and that a distinction can be made between strongly bonded and weakly bonded water molecules. Following this view, the two HP-DTA peaks may be related respectively to the expulsion of the weakly bonded water (low-temperature peak) and, after the hydration shells of the interlayer cations have been reduced, to a secondary water-loss (high-temperature peak). The latter event probably involved the inner-hydration shell of the interlayer cation and was dependent not only on bonding effects of the interlayer cation, but also on interactions with the 2:1 layer. Because in these experiments the chemistry of the 2:1 layer was the same, the different results must be related to the nature of the interlayer cation. This effect is shown on the P-T diagrams in Figures 9 and 10.

Also shown in Figures 9 and 10 is the reaction $(L,V)_{H_2O}$. The dehydration reactions of various montmorillonite samples followed the general shape of $(L,V)_{H_2O}$, but the dehydration reactions in several runs occurred at anomalously low temperatures. These reactions probably represented conditions at which $P_{H_2O} < P_{total}$. The near-horizontal slope of the curves at low pressures explains the previously described observations regarding the important effect of small differences in γ_{H_2O} in thermal analysis. The temperatures and curvatures of the first dehydration peak were different for the different cations, whereas the temperatures of the second dehydration reaction were less affected by the interlayer cation type.

Determination of the enthalpy of a reaction on the basis of the area of a peak at a single pressure is possible if all variables involving the reaction can be evaluated accurately. This determination is very difficult because the calibration of the instrument may vary with temperature, reaction rate, and the size and thermal conductivity

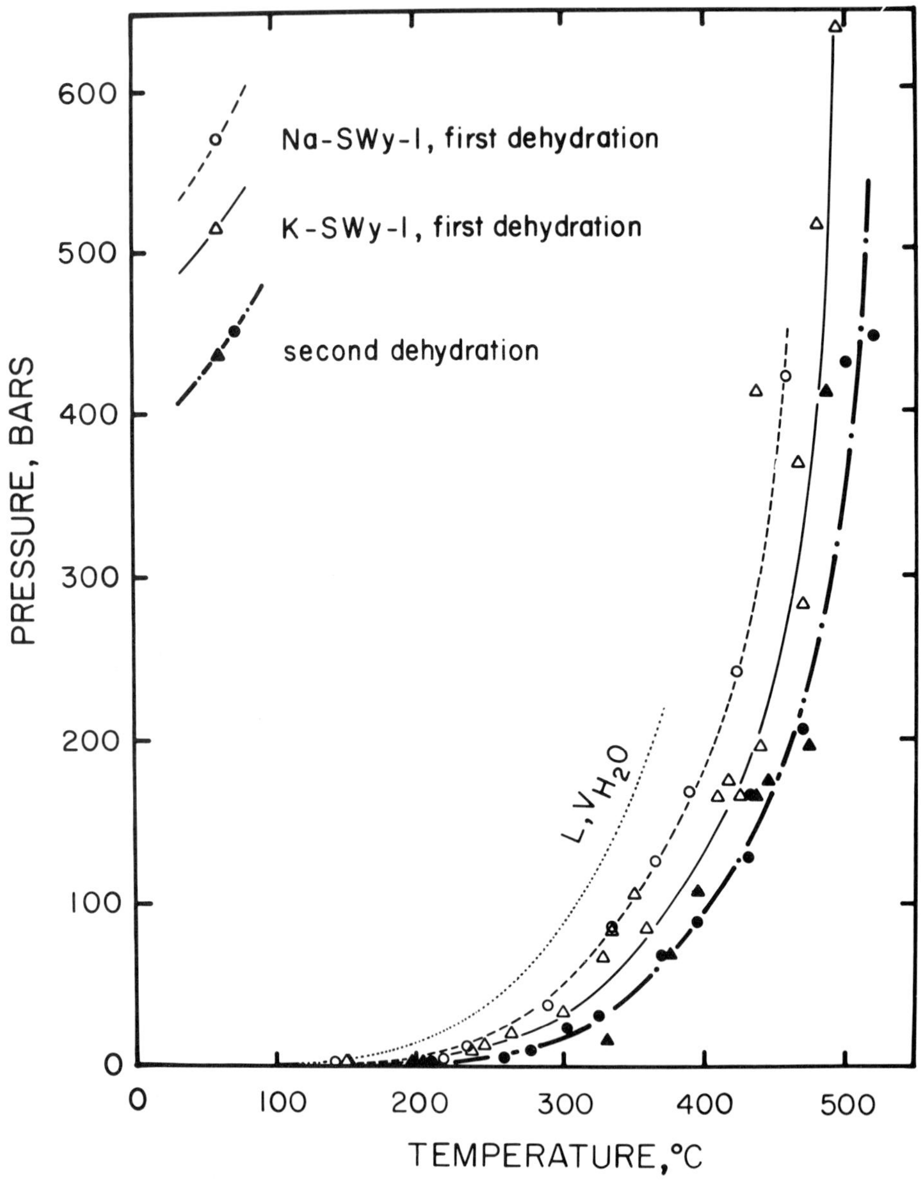

Figure 9. P-T Projections of dehydration reactions of samples SWy-1 and K-SWy-1. Reaction $(L,V)_{H_2O}$ is shown for comparison. Open symbols represent first dehydration reactions; solid symbols represent second dehydration reactions (Koster van Groos and Guggenheim, 1986).

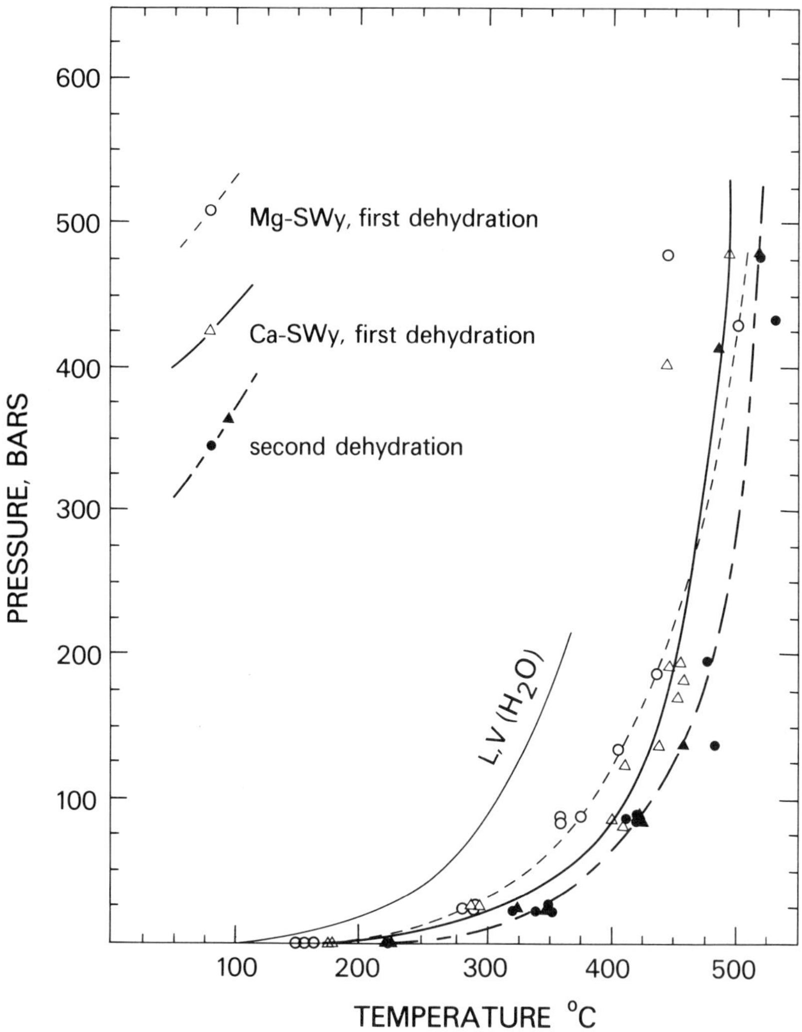

Figure 10. P-T projections of dehydration reactions of samples Ca-SWy-1 and Mg-SWy-1. Reaction $(L,V)_{H_2O}$ is shown for comparison. Open symbols represent first dehydration reactions; solid symbols represent second dehydration reactions (Koster van Groos and Guggenheim, 1987a).

of the sample. Calorimetric methods using the heat of formation of the reactants and products are generally more precise. In fine-grained, natural material, such as clay, however, the accuracy of calorimetric methods may be adversely affected by impurities (e.g., colloidal quartz). A fundamentally different approach is to determine the enthalpy of reaction from the P-T relation of a reaction, using $dP/dT = \Delta H/T\Delta V$ (Clapeyron equation). If ΔV can be evaluated, the reaction enthalpy may be calculated.

If a gas evolves in a reaction, the difference in the molar volume of the solids relative to the gas phase is small, and, therefore, the effect of the solid phase on V may be ignored. The enthalpy of this type of reaction may be evaluated using the relation:

$$\ln P_2 - \ln P_1 = (\Delta H/R)(1/T_2 - 1/T_1)$$

(see, e.g., Anderson, 1977). This relationship is only valid if the fugacity coefficient $\gamma_{H_2O} = 1$ and the heat capacity, C_p, approaches zero. For example, a plot of $(L,V)_{H_2O}$ using data from the Steam Tables (Keenan *et al.*, 1978), shows that although γ_{H_2O} ranges considerably over the P-T range, the plotted data lie on a straight line (Figures 11 and 12). The slope of this line yields for the reaction $(L,V)_{H_2O}$ a ΔH value of 39.5 kJ/mole at 1 bar and 100°C. If a 2% adjustment is made (because $\gamma_{H_2O} = 1$ was assumed rather than $\gamma_{H_2O} = 0.98$ (Helgeson and Kirkland, 1974)), the 39.5 kJ/mole value changes to 40.29 kJ/mole, which is very close to the enthalpy of evaporation determined by calorimetry (cf. 40.886 kJ/mole, Robie *et al.*, 1978). These results suggest the utility of such a plot for the determination of the enthalpy of the dehydration reaction.

The enthalpy of dehydration at 1 bar, ΔH_{dh}, for interlayer water was obtained for each montmorillonite (see Figures 11 and 12). It is important in deriving ΔH_{dh} that $P_{H_2O} = P_{total}$, otherwise the dehydration reaction temperatures are lowered. Critical assessment of the resulting data is, therefore, necessary. In these studies, greater weight should be given to the 1-bar data and to the results yielding the higher dehydration temperatures, because these data more likely represent the condition $P_{H_2O} = P_{total}$. Estimating standard deviations for the derived values should take the uncertainty of the experimental conditions into account.

The enthalpy of a reaction is the total heat involved in the breaking and forming of bonds. The dehydration reaction of montmorillonite involves only the loss of interlayer water, and the enthalpy of dehydration, ΔH_{dh}, may be thought of as the sum of the enthalpy of the interlayer water, ΔH_{iw}, involving the bonding effects and H_{ev}, the enthalpy of evaporation of water. Thus, the enthalpy of interlayer water at 1 bar, ΔH_{iw}, may be derived from the relationship $\Delta H_{dh} = H_{iw} + H_{ev}$. The enthalpy of evaporation, ΔH_{ev}, can be determined by extrapolating data from $(L,V)_{H_2O}$ (Keenan *et al.*, 1978) at 1 bar and 100°C to the temperature of dehydration.

The enthalpy values for dehydration and for interlayer water are given in Table 2. ΔH_{iw} values for the (low-temperature) dehydration event differ only slightly for the four compositions; however, values for monovalent interlayer cations are more alike than are values for divalent interlayer cations. Note that the temperature of the dehydration reaction depends on the stability of the hydrated phase and not only on the enthalpy of dehydration. These relationships appear reasonable because interlayer cations should have little influence on the voluminous and weakly bonded second and third nearest-neighbor H_2O molecules (i.e., the "outer hydration shell").

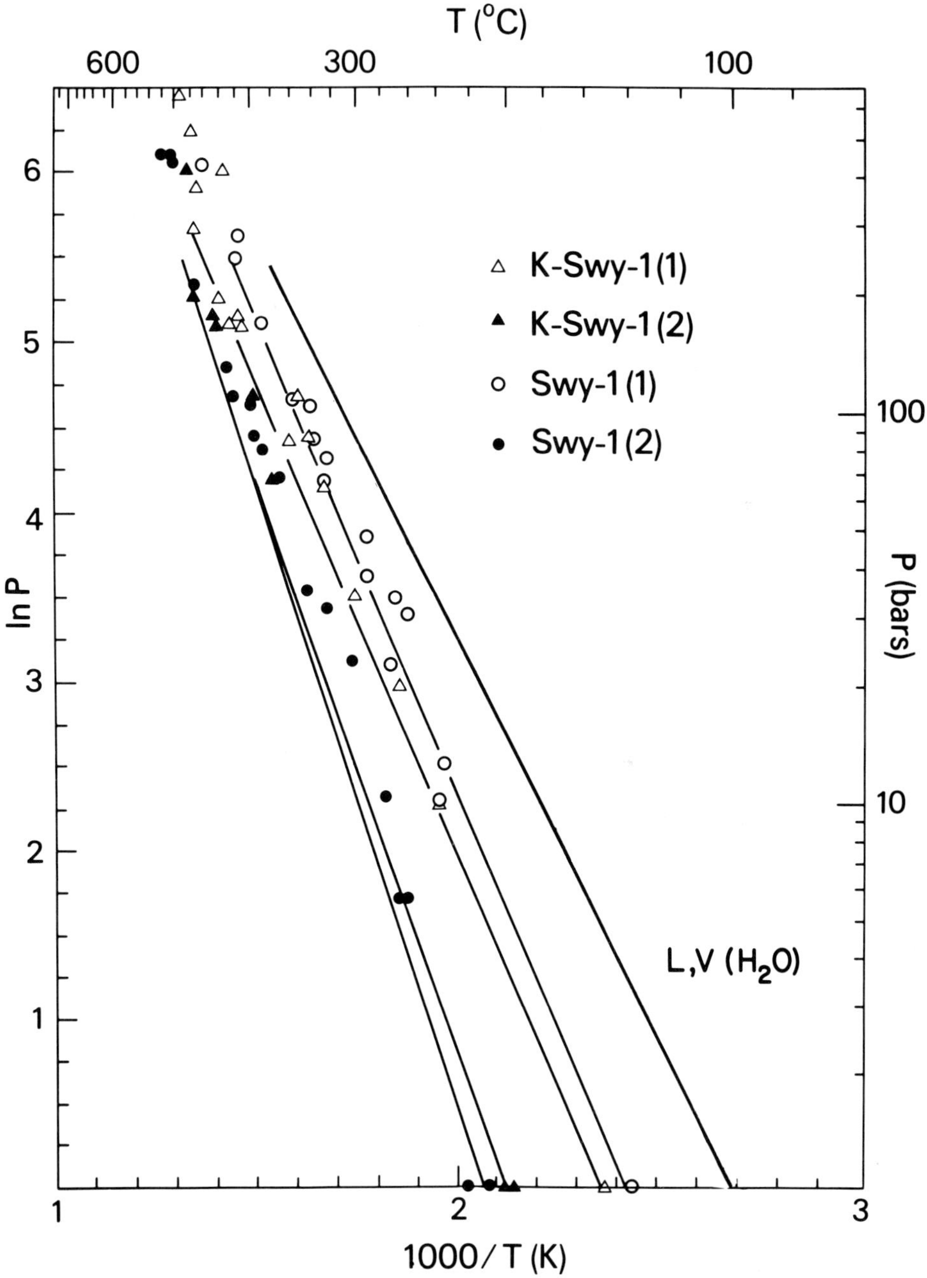

Figure 11. ln P-1000/T projection of dehydration reactions of samples SWy-1 and K-SWy-1. Reaction $(L,V)_{H_2O}$ is taken from Keenan *et al.* (1978). (Koster van Groos and Guggenheim, 1986).

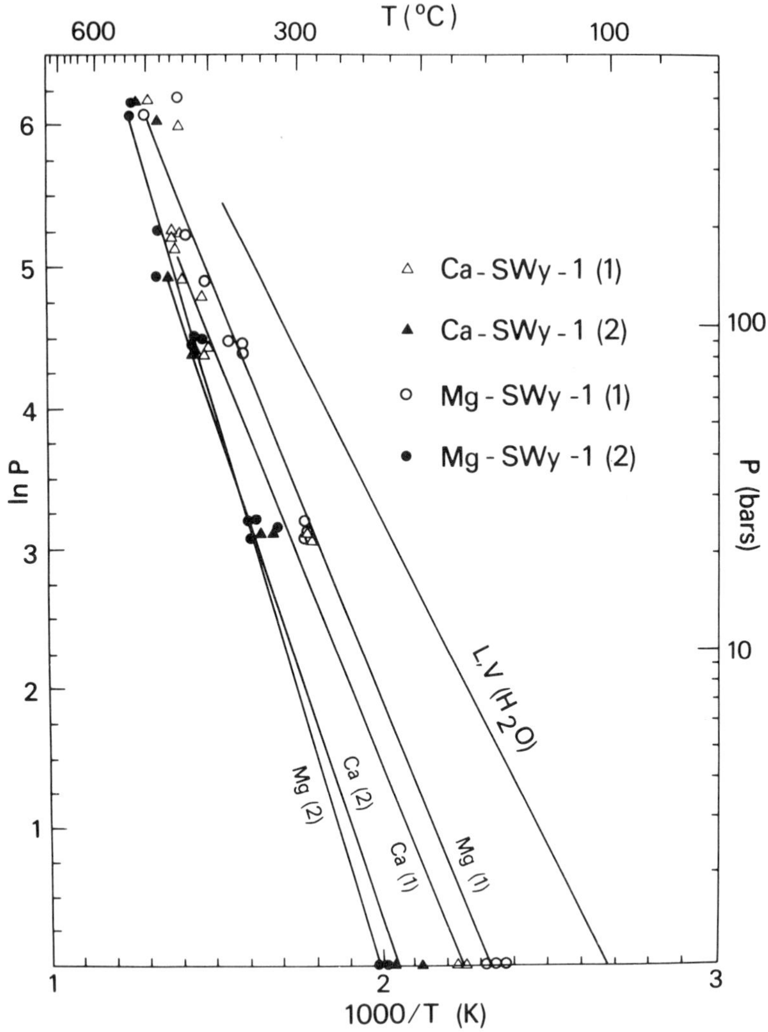

Figure 12. ln P-1000/T projection of dehyration reactions of samples Ca-SWy-1 and Mg-SWy-1. Reaction $(L,V)_{H_2O}$ is taken from Keenan *et al.* (1978). (Koster van Groos and Guggenheim, 1987a).

The higher-temperature, second dehydration reaction involves more tightly bound H_2O ("inner hydration shell"), which is influenced more by the type of interlayer cation and the effects of charge-undersaturated, basal oxygens on the surface of the 2:1 layer (for any tetrahedra with Al substituting for Si). The 2:1 layer, however, was kept constant in all experiments; therefore, the differences in H_{iw} for the second reaction can be attributed to the effect of the interlayer cation only. Thus, the thermally induced behavior of the inner hydration shell of H_2O molecules is related to both size and charge of the interlayer cation.

The data reported in Figures 7 and 8 confirm the observations that the enthalpy of the second dehydration of sample K-SWy-1 is much lower than that of sample Mg-SWy-1. Both the similarities between the enthalpy of the first dehydration and the differences of the second dehydration of these samples indicate that the P-T curve of their reactions, as shown in Figures 9 and 10, cannot be drawn accurately without considering the thermodynamic implications.

Results from thermal gravimetric analyses (TGA) (Koster van Groos and Guggenheim, 1987a) are summarized in Table 3. Because the starting material was obtained by purification and size-fractionation of montmorillonite SWy-1 without homoionic exchange with Na, its H_2O content may have been different from Na-exchanged SWy-1. Other values relating to sample SWy-1 are based on TGA. The enthalpy (Table 2) and associated number of H_2O molecules per cation in the interlayer (Table 3) can be used to establish a physical model consistent with both.

Physical model. Extremes in the value of Δ_{Hiw} for the higher-temperature (second) dehydration reaction (cf. samples K-SWy-1: ΔH_{iw} = 19.8 kJ/mole; Mg-SWy-1: ΔH_{iw} = 31.1 kJ/mole) suggest that a discussion of the physical significance of the data should emphasize the results from samples K-SWy-1 and Mg-SWy-1. The large value of ΔH_{iw} for sample Mg-SWy-1 indicates that the remaining water molecules

Table 2. Enthalpy (kJ/mole) of dehydration, ΔH_{dh}, of first and second dehydration reactions and enthalpy of interlayer water, ΔH_{iw}, of smectite samples.

	Dehydration	ΔH_{dh}[1]	ΔH_{dh}[2]	ΔH_{dh}[3]	ΔH_{iw}
SWy-1	First	46.45	47.1	46.8 ± 0.5	7.8 ± 0.5
	Second	61.14	64.7	62.9 ± 2	27.1 ± 2
K-SWy-1	First	46.1	46.22	46.2 ± 0.5	7.8 ± 0.5
	Second	54.86	58.64	56.7 ± 2	19.8 ± 2
Ca-SWy-1	First	49.4	49.59	49.5 ± 0.51	2.1 ± 0.5
	Second	57.42	59.12	58.2 ± 2	22.6 ± 2
Mg-SWy-1	First	49.12	49.56	49.3 ± 0.5	11.1 ± 0.5
	Second	66.88	65.57	66.2 ± 2	31.1 ± 2

[1]Taken from two data points.
[2]Using least squares analysis.
[3]Obtained by averaging.

Table 3. Estimated number of water molecules per interlayer cation.[1]

Cation	Total (wt. %)	Total H_2O molecules	Est. distribution 1st react.	2nd react.
Na	9	4-5	4-5	0.5
K	6	3-4	3-4	«0.5
Ca	16	13	12	1
Mg	16	13-14	11	2

[1]Based on thermal gravimetric analyses.

were bound to the interlayer (involving the effects of the interlayer cation and basal plane surface) more tightly than weaker molecules associated with the other interlayer cations. This interpretation is consistent with the relatively high temperature of the reaction (≈225°C for sample Mg-SWy 1 vs. ≈215°C (Na), ≈210°C (Ca), ≈195°C (K)), suggesting that part of the Mg-H_2O complex more easily enters the silicate rings of the 2:1 layer, thus bonding water molecules more tightly to the structure. Because H_2O is polar, this arrangement should also help compensate for the undersaturated nature of basal oxygens associated with Al-substituted tetrahedra. Larger interlayer cations are more effective in preventing water molecules from entering the silicate rings; K prevents essentially all water molecules from entering the ring.

Application to geologic systems. Understanding the importance of varying H_2O fugacity in these experiments allows the effect of lower H_2O activity in sediments to be evaluated. In sedimentary sequences the H_2O activity can be reduced in several ways. For example, an increase in the concentration of solutes, such as salts or perhaps hydrocarbons, in ground water or a reduction of the hydrostatic pressure with respect to the load pressure substantially reduces the activity of water. Such a reduction should lower the dehydration temperature of montmorillonite and, therefore, montmorillonite may dehydrate at the conditions found in sedimentary strata, thus buffering the activity of water. This sequence of reactions could result in a rapid dehydration event, should the pressure not be maintained during drilling in montmorillonite-bearing strata.

Of equal importance is the observation that at high pressures the interlayer water of smectite is stable ≥400°C. This has an important bearing on the water content of the lower crust or upper mantle at destructive plate margins. Thus, a down-thrusted slab, which is relatively cold and in which fluid circulation is limited, should be able to transport substantial amounts of water present in the interlayer of smectite to considerable depth. This discussion assumes, of course, that composition is fixed and that the smectite does not alter to illite or to another more stable phase assemblage.

Dehydroxylation reactions (dioctahedral aluminous phases)

During dehydration, relatively weakly bonded water is removed from the phyllosilicate structure, as was discussed in the previous section. The hydroxyl groups in these structures are much more strongly bonded, and higher temperatures are necessary for their removal. The temperature of the DTA peak or peaks representing dehydroxylation is related to the nature of the hydroxylated phase and, thus, to its structure and chemistry. For example, the kaolinite structure (1:1 type layer) differs considerably from the montmorillonite structure (2:1 type layer) in that most of the hydroxyl groups in kaolinite are located in the plane adjacent to the interlayer, whereas in montmorillonite they are within the 2:1 layer. This may be related to the lower dehydroxylation temperature (≃600°at 1 bar) of kaolinite compared with montmorillonite (≃700°C at 1 bar), although both phases are aluminuous dioctahedral materials.

In general terms, the delocalization mechanism of the hydrogen must be sufficiently similar in these phases to dominate the phase relations. More specifically, in the kaolinite-H_2O and montmorillonite-H_2O systems, dehydroxylation involves a topotactic transformation at lower pressure (≃45 bars), with the formation of metakaolinite and montmorillonite dehydroxylate, respectively.

Topotactic reactions are of particular interest because they may yield a product having sufficient structural continuity such that the resulting phase may be studied by single crystal X-ray diffraction techniques. Complete understanding of a reaction must include models at the atomic level, as well as thermodynamic or macroscopic models. Pyrophyllite and muscovite topotactically dehydroxylate and remain structurally continuous throughout a single grain above the dehydroxylation temperature; they therefore produce suitable material for study so that models may be developed. Fortunately, these phases are sufficiently similar both structurally and chemically to montmorillonite that they offer an analogue system of study. A similar analogue system for kaolinite is unknown, but understanding the dehydroxylation mechanism of the aluminous 2:1 dioctahedral phyllosilicates provides insight into the thermal decomposition process in general.

Physical model for dehydroxylation in 2:1 layers. Guggenheim *et al.* (1987) suggested a model for the dehydroxylation of 2:1 phyllosilicates in which the strength of the Al-OH bond is greatly affected by the coordination number of neighboring cations. This model, developed from first principles by using Pauling bond-strength summations, may be used to explain the shape of peaks derived from thermo-analytical experiments. The model, illustrated by pyrophyllite, is presented in Figure 13.

Pauling's electrostatic valency principle states that in a stable structure, the total strength of the bonds that reach an anion from all neighboring cations is equal to the formal charge of the anion, but opposite in sign. Figure 13A shows Pauling bond-strength summations for selected oxygen sites in pyrophyllite; Figure 13B shows similar summations for the pyrophyllite dehydroxylate structure, i.e., the high-temperature phase that forms topotactically from pyrophyllite. Unlike pyrophyllite, in which all oxygens are charge-saturated at 2.0 electrostatic valency units (esu), the dehydroxylate consists of Al polyhedra containing one undersaturated oxygen (at 1.2 esu) and the remaining oxygens oversaturated with respect to charge (at 2.2 esu undersaturated oxygen and away from the over-saturated oxygens. This qualitative prediction of changes in bond distances is important, because bond distance is directly correlated to bond strength, with smaller distances corresponding to stronger bonds.

Similar summation calculations may be made for a partly dehydroxylated pyrophyllite structure (Figure 13C). Note here that some Al cations are in five-fold coordination as in the dehydroxylate structure and some are in six-fold coordination as in pyrophyllite. Calculated summation values show asymmetric distributions for each six-fold coordinated Al cation. Therefore, this cation moves away from the side of the octahedron containing oversaturated oxygens and towards the saturated oxygen of the hydroxyl group. Hydroxyl protons can readjust their positions to polarize further the hydroxyl oxygen, thereby making a closer approach of the Al cation to the hydroxyl more stable. The increased attraction between the Al and its coordinating hydroxyl is a consequence of the neighboring five-fold coordination site.

The model predicts, therefore, partial dehydroxylation at lower temperatures for structures in which Al cations adjacent to the pair of Al octahedra are in six-fold coordination. After partial dehydroxylation, however, the remaining hydroxyls are more tightly held if neighboring Al cations are in five-fold coordination. A bimodal distribution of hydroxyl loss is anticipated (Figure 14). Both five-fold and six-fold Al neighbors are possible; such arrangements produce broadening of the thermal peaks. In addition, because hydroxyl (or H_2O) diffuses through the interlayer (see Guggenheim *et al.*, 1987, for additional details), interlayer site occupancy also affects the shape of the thermal peak and the temperature of dehydroxylation.

Dehydroxylation studies using HP-DTA have been directed at understanding the effect of interlayer cation chemistry and, consequently, the number of interlayer sites occupied. As with the parallel dehydration studies (see above), Koster van Groos and Guggenheim, 1987b, 1989) reported HP-DTA experiments for samples SWy-1, K-SWy-1, Mg-SWy-1, and Ca-SWy-1. Because the cations were exchanged under ambient conditions, the 2:1 layer chemistry was unmodified. Thus, dehydroxylation could be investigated as a function of differences of the interlayer region.

Differential thermal gravimetric (DTG) curves for montmorillonite samples (Figure 15) show a broad asymmetric peak above 400°C, with the asymmetry being more pronounced for the Na phase. By analogy with pyrophyllite and muscovite (Guggenheim *et al.*, 1987), the broad peak probably represents a major peak at

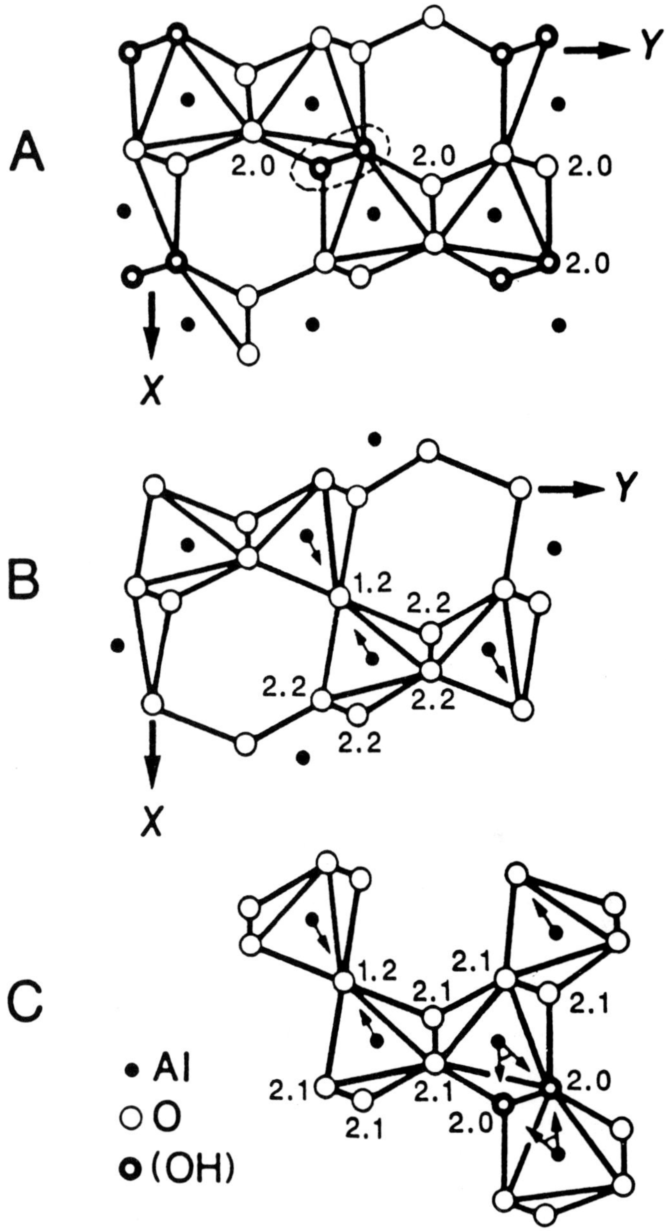

Figure 13. Al polyhedra of (A) pyrophyllite, (B) pyrophyllite dehydroxylate and (C) pyrophyllite in transition due to thermal stress. Numbers associated with oxygen atoms refer to Pauling bond-strength summations (see text). Dashed oval in (A) refers to the pair of hydroxyls involved in water loss [cf. (B) and (C)]. Arrows associated with Al cations indicate predicted direction of movement as determined from summation calculations (from Evans and Guggenheim, 1988, as modified from Guggeneheim *et al.*, 1987).

700°C and a smaller, shoulder peak at 650°C for sample SWy-1, and a major peak at 670°C and a lesser peak at 650°C for sample K-SWy- 1. Thus, the temperature of dehydroxylation of sample K-SWy-1 is lower than that of sample SWy-1, although the onset temperature for these reactions is identical. The shoulder dehydroxylation peak at the lower temperature side may be related to the loss of OH, where neighboring polyhedra are in six-fold coordination. Because OH may be lost from distributions that include both five-fold and six-fold polyhedral neighbors, the thermal peaks are broad. The TGA and DTG curves for the divalent samples Ca-SWy-1 and Mg-SWy-1 at high temperatures (>500°C) are similar; the range in dehydroxylation temperature is about the same (575°-750°C for Na, 600°-775°C for Mg, Ca), although the temperatures should not be compared closely because H_2O fugacities may differ slightly (Koster van Groos and Guggenheim, 1989). A major difference is that the DTG curves are more asymmetric, with the curve of sample Mg-SWy-1 being resolved into two peaks.

Interpretation of HP-DTA data. Figure 16 shows the HP-DTA signals of selected, open-capsule analyses of samples SWy-1, K-SWy- 1, Ca-SWy-1, and Mg-SWy-1; Figure 17 shows the open-capsule results for kaolinite. Figures 18 and 19 show closed-capsule results for kaolinite and montmorillonite samples SWy-1, and K-SWy-1 (Yeskis *et al.*, 1985; Koster van Groos and Guggenheim, 1987b, 1989). The peak shape at lower pressures (<45 bars) differs from the higher-pressure experiments with these materials, which suggests that the dehydroxylation mechanism was affected by pressure. For example, at 1 bar the peak width for the kaolinite dehydroxylation reaction observed in open-capsule experiments was ≃100°C compared with ≃20°C at 1240 bars. A similar reduction in peak width was observed for sample SWy-1, more so for sample K-SWy-1, and especially for sample Ca-SWy-1; sample Mg-SWy-1 behaved differently in that two distinct dehydroxylation reactions were present, which become increasingly resolved at higher pressures (note labels at about 740°C and 790°C in Figure 16D).

These narrow peaks indicate that the increase in pressure either affected the dehydroxylation process by making the OH groups more alike energetically, resulting in a more uniform dehydroxylation event, or that a different type of reaction occurred. This effect may have been augmented by an increase in the rate of the dehydroxylation due to the increase in temperature. In contrast, in closed-capsule experiments in which $P_{H_2O} = P_{total}$, the peaks of kaolinite (Figure 18) were only slightly sharper at higher pressures, whereas in experiments with montmorillonite (Figure 19) no reduction in peak-width was observed at higher pressures (Koster van Groos and Guggenheim, 1987b, 1989). The differences in the DTA patterns of open- and closed-capsule experiments strongly suggest that different reactions were involved.

Results and discussion - kaolinite. The phase rule relationships discussed below represent metastable equilibrium (Roy and Osborn, 1954). The stable reaction, kaolinite = hydralsite + montmorillonite + pyrophyllite, occurs at about 400°C over a range of pressure. The dehydroxylation reaction observed in open capsules for

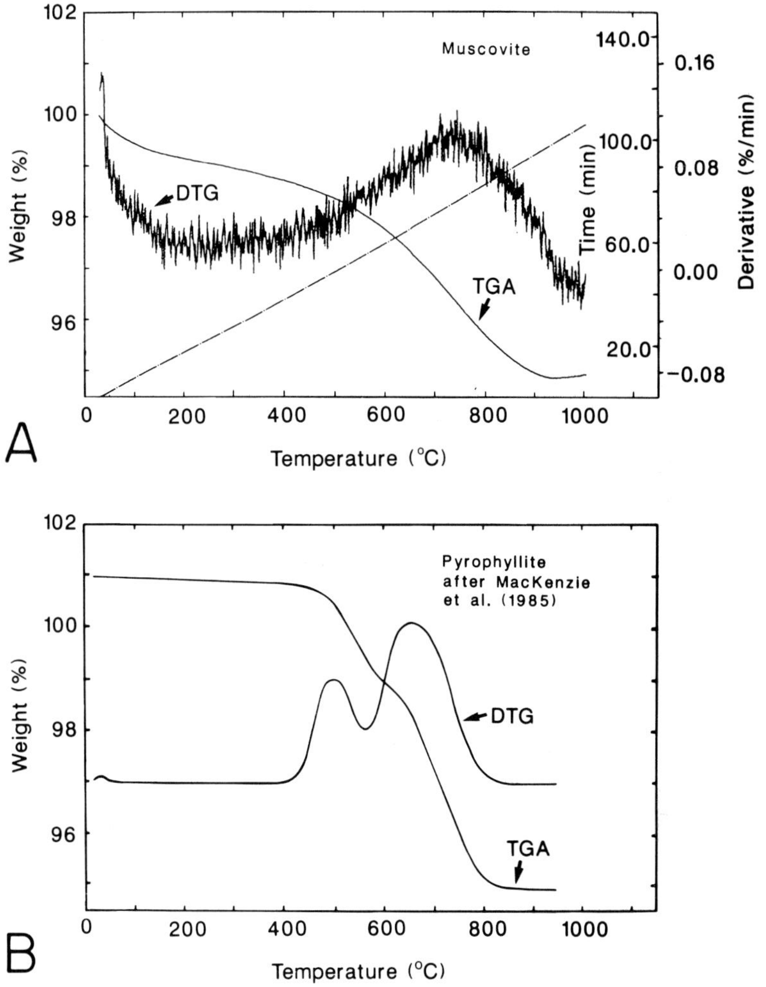

Figure 14. Thermal gravimetric and differential thermal gravimetric analysis patterns for (A) muscovite and (B) pyrophyllite (from Mackenzie *et al.*, 1985; Guggeheim *et al.*, 1987).

kaolinite (curve I, Figure 20) may be divided into two parts, I_a and I_b, on the basis of the effect of pressure. The portion of the curve labeled I_a is greatly dependent on pressure, whereas I_b is not, suggesting that the two parts of curve I involved different reactions. Reaction I_a involves a large volume change, which must be related to the evolution of a vapor phase. Evidence from X-ray powder diffraction data suggests that this reaction is kaolinite (K) = metakaolinite (MK) + vapor (V), where metakaolinite is an ill-defined kaolinite-like structure containing, perhaps, some residual

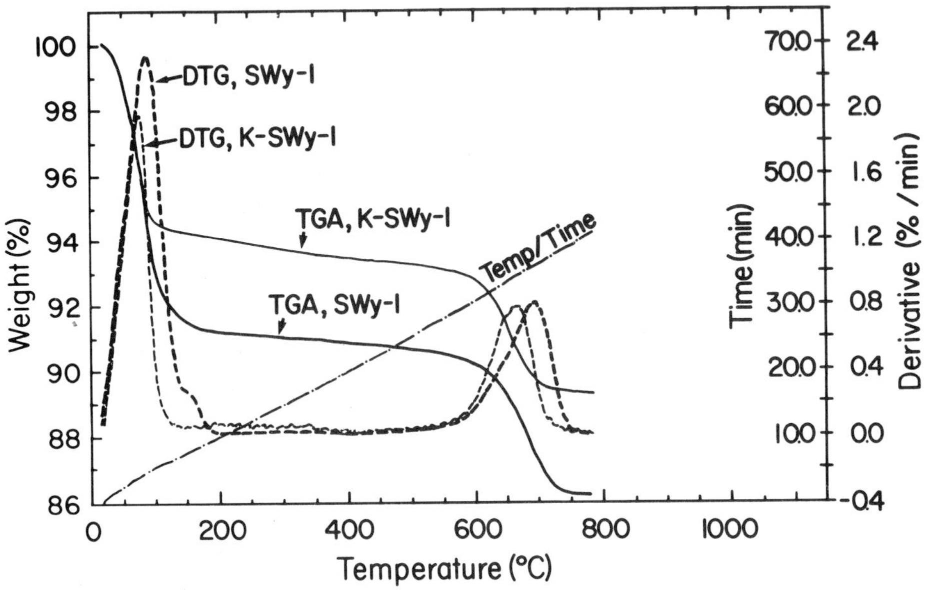

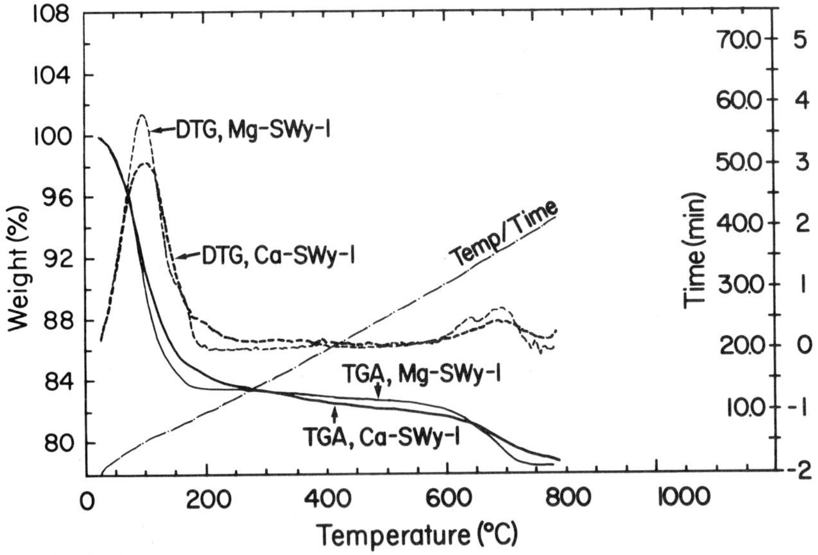

Figure 15. Thermal gravimetric and differential thermal gravimetric analysis patterns for (upper) samples SWy-1 and K-SWy-1 and (lower) samples Ca-SWy-1 and Mg-SWy-1. Note that interlayer water content of sample K-SWy-1 is much less than for sample SWy-1 (Koster van Groos and Guggenheim, 1987b, 1989).

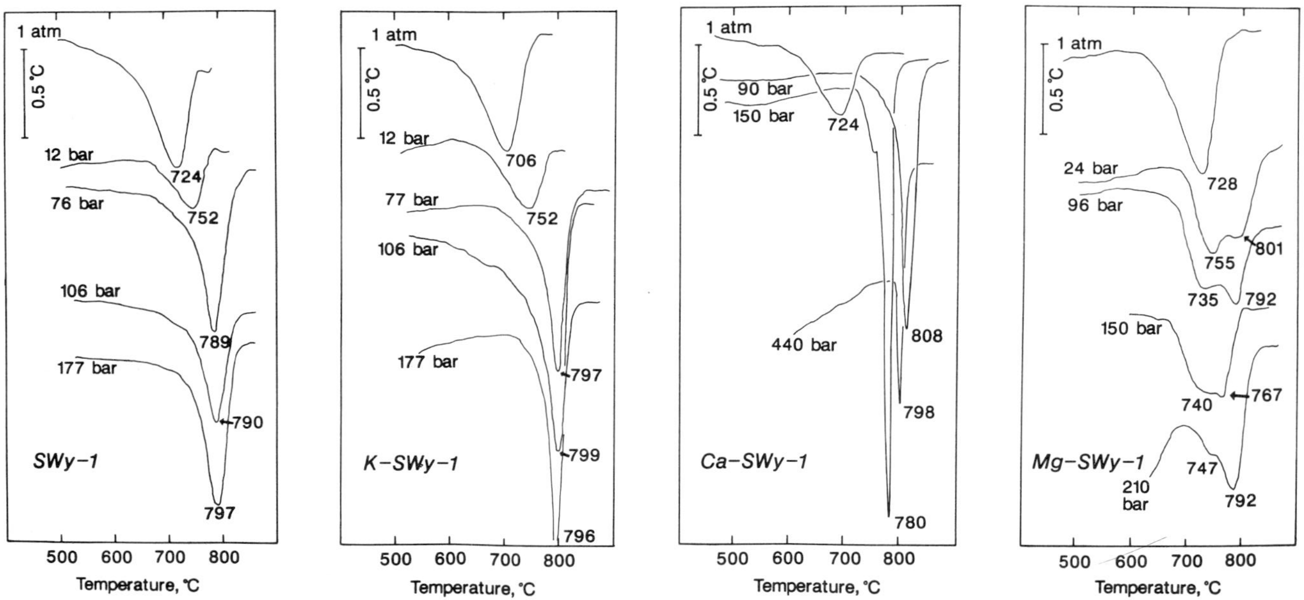

Figure 16. High-pressure differential thermal analysis patterns of dehydroxylation reactions of (A) sample SWy-1, (B) sample K-SWy-1, (C) sample Ca-SWy-1, and (D) sample Mg-SWy-1, using open capsules. Q = quartz. (From Koster van Groos and Guggenheim, 1987b, 1989).

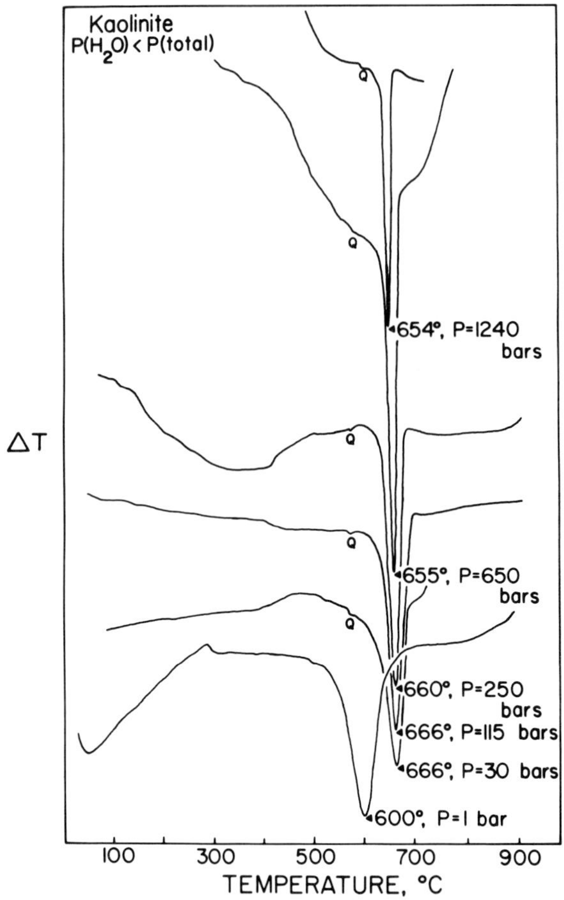

Figure 17. High-pressure differential thermal analysis patterns of kaolinite KGa-1 for open-capsule runs without added water. Q = quartz. (From Yeskis *et al.*, 1985).

hydroxyl (Roy *et al.*, 1958; Brindley and Nakahira, 1959). The evolving vapor phase was probably pure H_2O. In contrast, reaction I_b was almost pressure independent, which suggests a small volume change; hence, a vapor phase is not likely to take part in this reaction.

Reaction II was observed in closed-capsule experiments and must have involved an assemblage containing both kaolinite and water vapor. The slope of the reaction curve is negative to about 2-3 kbars, above which it becomes positive. These slopes suggest that a liquid-like phase having a variable H_2O content was produced, because the pronounced negative slope of this reaction is typical of solid + vapor = liquid reactions. Yeskis *et al.* (1985) called this liquid-like phase "metaliquid" (ML). Reaction II, therefore, can be written as kaolinite + vapor = ML. The formation of the ML is supported by the lack of a substantial amount of crystalline

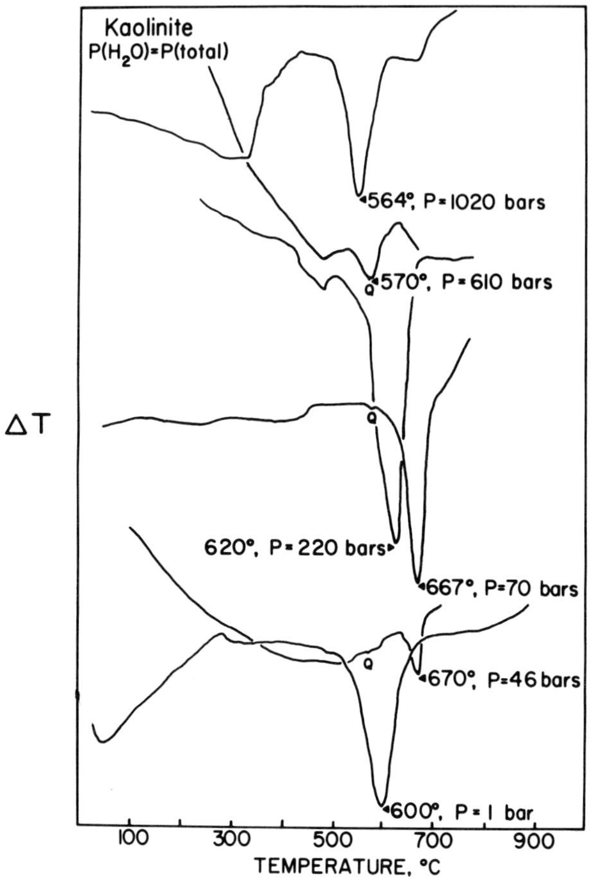

Figure 18. High-pressure differential thermal analysis patterns of kaolinite (KGa-1) for closed capsule runs with added water. Q = quartz. (From Yeskis *et al.*, 1985.)

phases in the run products, with the exception of quartz that was present in the starting material and a small amount of an unknown, weakly diffracting phase (ML or its quench products. The ML phase may be thought of as a poorly crystalline or noncrystalline phase, having a variable H_2O content and the same Al/Si ratio as kaolinite.

The formation of a ML is unusual, and the ML may be very similar to metakaolinite. Substantial evidence exists, however, against equating these two phases. A metakaolinite structure cannot accommodate large variations in OH content. Furthermore, at about 40 bars, Yeskis *et al.* (1985) encountered reaction II at ≈20°C higher than reaction I. The arguments for postulating the ML phase are (1) the glassy nature of the run products, (2) the absence of a substantial amount of a crystalline phase, (3) the nature of the slope of reaction curve II, and (4) other arguments based

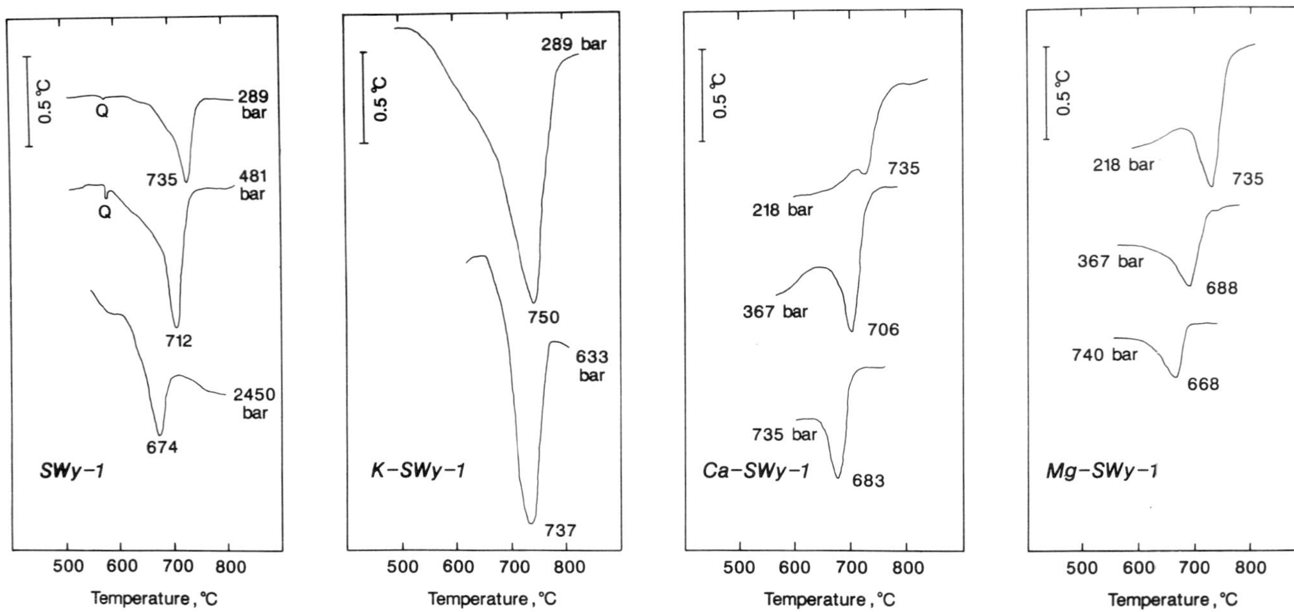

Figure 19. High-pressure differential thermal analysis patterns of dehydroxylation reactions of (A) sample SWy-1, (B) sample K-SWy-1,(C) sample Ca-SWy-1, and (D) sample Mg-SWy-1, using closed capsules. Q = quartz. (From Koster van Groos and Guggenheim, 1987b, 1989).

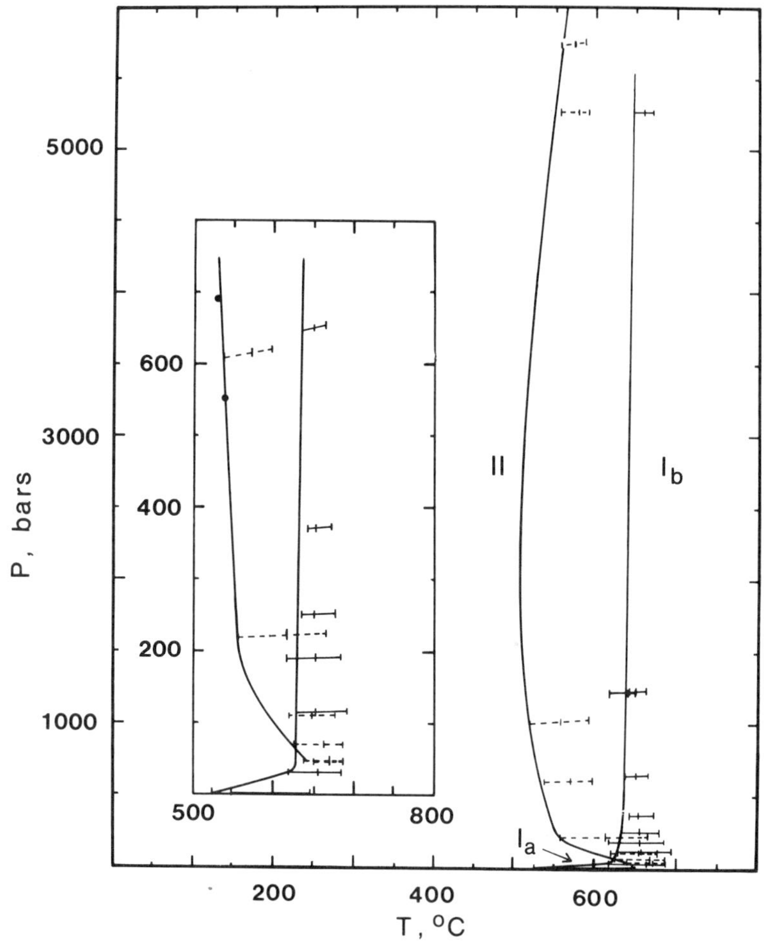

Figure 20. P-T Projections of dehydroxylation reaction of kaolinite. Inset shows details at low pressure. Bars show extrapolated onset, peak, and extrapolated return temperatures. Solid bars and dashed bars represent open-capsule and closed-capsule runs, respectively. Curves are drawn through onset temperatures: I_a = open capsules at low pressures; I_b = open capsules at high pressures; II = closed capsules; (Yeskis *et al.*, 1985). Solid circles (see inset) are data from Weber and Roy (1965).

on the phase relations, which are discussed below.

The above discussion indicates that in the kaolinite-H_2O system four phases are present: kaolinite (K), metakaolinite (MK), metaliquid (ML), and vapor (V). Roy *et al.* (1958) suggested the formation of γ-Al_2O_3 in these reactions. Likewise, for some kaolinite samples, mullite has been reported in reaction products along with γ-Al_2O_3 (Tsuzuki, 1961). Assemblages that contain these additional phases are probably transitional to equilibrium assemblages. As such, they are outside the scope of these reactions and need not be discussed here. Thus, the phase relations indicate that the system behavior is binary; the possible reactions are given in Table 4.

ML contains less, an equal amount, or more H_2O than kaolinite. For a water-poor ML, a condition that can be expected at low pressures, reactions 2a, 4a, and 6a are possible. With higher H_2O content (at higher pressures), reaction 2b can occur. Reactions 4b and 6b are impossible, as can be derived on the basis of Schreinemaker's rules and compositional considerations. Reaction 3 is a special case in which ML has the same H_2O content as kaolinite. Reaction 7 represents the melting of MK in the absence of vapor. It occurs at 1 bar at ≈850°C (Roy *et al.*, 1958, p. 204), and it should have a steep positive slope. Reaction 6a is invariant (see Figure 21) if the system is binary; MK and H_2O are the compositional end-members of the system. Reactions 1, 2a, 4a, and 5 emanate from the invariant point (I) to form Schreinemaker's bundle. In reaction 2a, ML contains less H_2O than does kaolinite, but with increasing pressure more H_2O must dissolve into the metaliquid. A singular point (S) results if the ML contains the same amount of H_2O as kaolinite, and reaction 3 becomes possible, extending to higher pressures. Because no vapor phase is involved in this reaction, the pressure effect on the reaction temperature is expected to be small. At these higher pressures, however, ML may also dissolve more H_2O than the kaolinite, and reaction 2b takes place. The three reactions terminate at the singular point (Figure 21).

The data of Yeskis *et al.* (1985) and Stone and Rowland (1954) locate the invariant point at 25 ± 5 bars and 625 ± 5°C and the singular point at 40 ± 10 bars and 640 ± 10°C. Reaction 5 was not observed by Yeskis *et al.* (1985), presumably because at higher temperatures most of the vapor phase was no longer present in the capsule.

The phase relations in the system kaolinite-H_2O can be understood as the intersection of the dehydroxylation reaction of kaolinite (reaction 1) which has a positive slope in P-T space, and the melting reaction of metakaolinite in the presence of an aqueous vapor (reaction 5) which has a negative slope. At the intersection, the system becomes invariant (reaction 6a). Complications arise from the fact that the variable H_2O content of the ML requires an additional reaction (reaction 3), generating a singular point. This additional reaction is supported by runs that indicate reaction 2b occurring at higher temperatures than reaction 4a at low pressures and the reverse at high pressures (Figure 21).

Results and discussion - montmorillonite. The thermal stability of the phases participating in the water-loss reactions in the kaolinite-H_2O and montmorillonite-H_2O systems differ, but the phase relations appear to be analogous; the geometry and the topology of the phase diagrams are remarkably similar. An equivalent topology has been found in other hydroxide water systems, such as the system CaO-H_2O (Wyllie, 1959). Like the kaolinite-H_2O system, the montmorillonite-H_2O phase relations represent metastable equilibria. The phases in this system are montmorillonite (M),

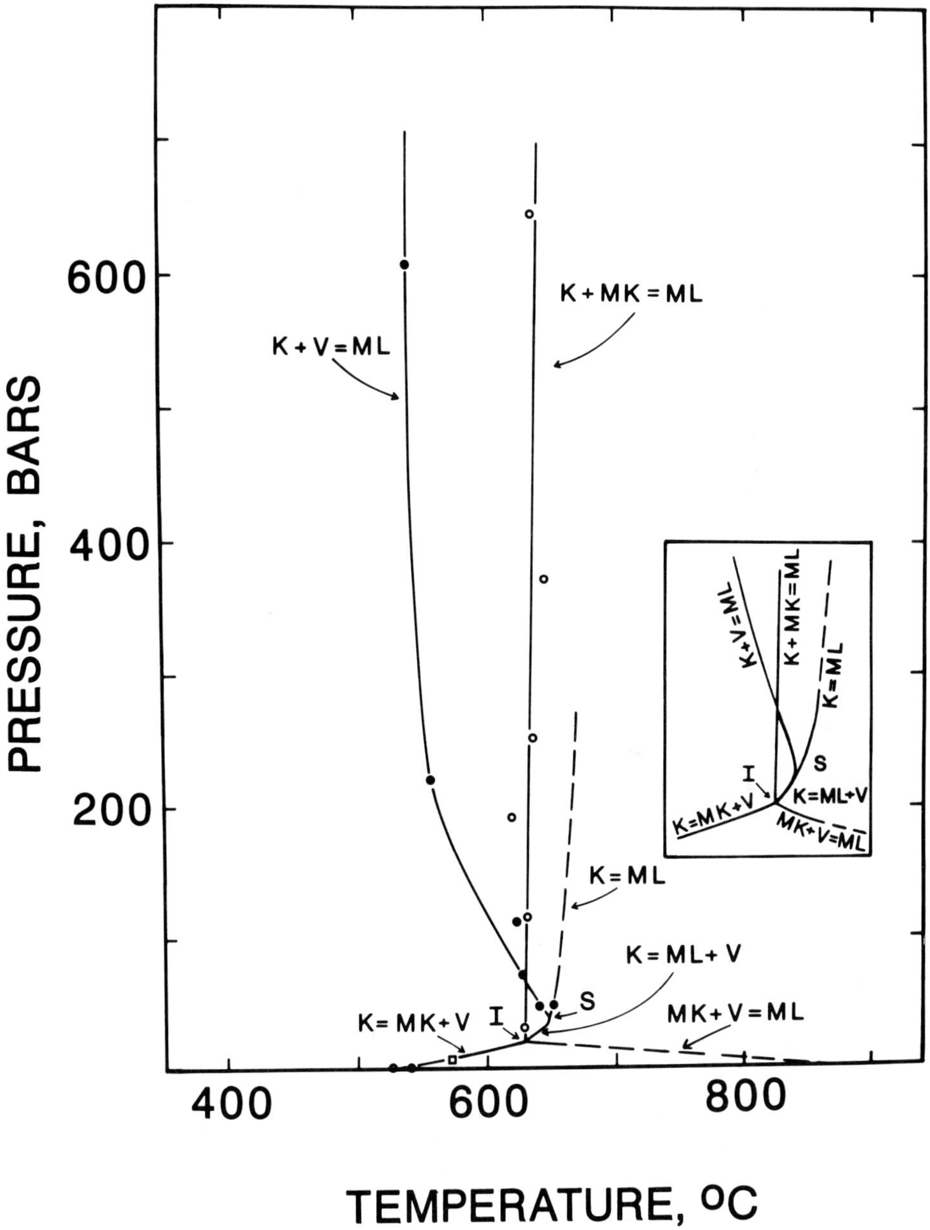

Figure 21. P-T relations in metakaolinite-H_2O system. Inset shows reactions at invariant points (Yeskis *et al.*, 1985). Open circles = open-capsule runs; solid circles = closed-capsule runs; open squares = data from Stone and Rowland (1954); K = kaolinite; MK = metakaolinite; ML = metaliquid; V = vapor; I = invariant point; S = singular point.

Table 4. Possible reactions in the system metakaolinite-H_2O.

Reactant		Product	Reaction number[1]	Reactant		Product	Reaction number[1]
K	=	K + V	1	K	=	MK + ML	4b
K	=	ML + V	2a	MK + V	=	ML	5
K + V	=	ML	2b	K + MK	=	ML + V	6a
K	=	ML	3	K + V	=	MK + ML	6b
K + MK	=	ML	4a	MK	=	ML	7

K = kaolinite; V = vapor; MK = metakaolinite; ML = metaliquid.
[1]See text.

Table 5. Possible reactions in the system montmorillonite-H_2O.

Reactant		Product	Reaction number[1]	Reactant		Product	Reaction number[1]
M	=	MD + V	1	M + MD	=	L	4
M	=	L + V	2a	MD + V	=	L	5
M + V	=	L	2b	M + MD	=	L	6
M	=	L	3	MD	=	L	7

M = montmorillonite; V = vapor; MD = montmorillonite dehydrate; L = liquid.
[1]See text.

montmorillonite dehydroxylate (MD), a liquid phase (L) having a variable H_2O content, and vapor (V). The existence of a montmorillonite dehydroxylate is based on X-ray powder diffraction data and by analogy with other aluminous dioctahedral 2:1 phyllosilicates (see above; Guggenheim *et al.*, 1987; Koster van Groos and Guggenheim, 1987b). Possible reactions, which are very similar to those in the kaolinite-H_2O system, are listed in Table 5. Reactions 6 represents an invariant point from which four univariant reactions, (i.e, 1, 2a, 4, and 5) emanate. Reaction 2a terminates in a singular point at which the H_2O content of the liquid is the same as in montmorillonite. From the singular point, reactions 2b and 3 extend to higher pressures, and the liquid phase is richer in H_2O or has the same H_2O content as montmorillonite, respectively. Reaction 5 represents the melting of MD in the presence of H_2O; it occurs at 1 bar at some temperature >900°C. Reaction 7 is the melting

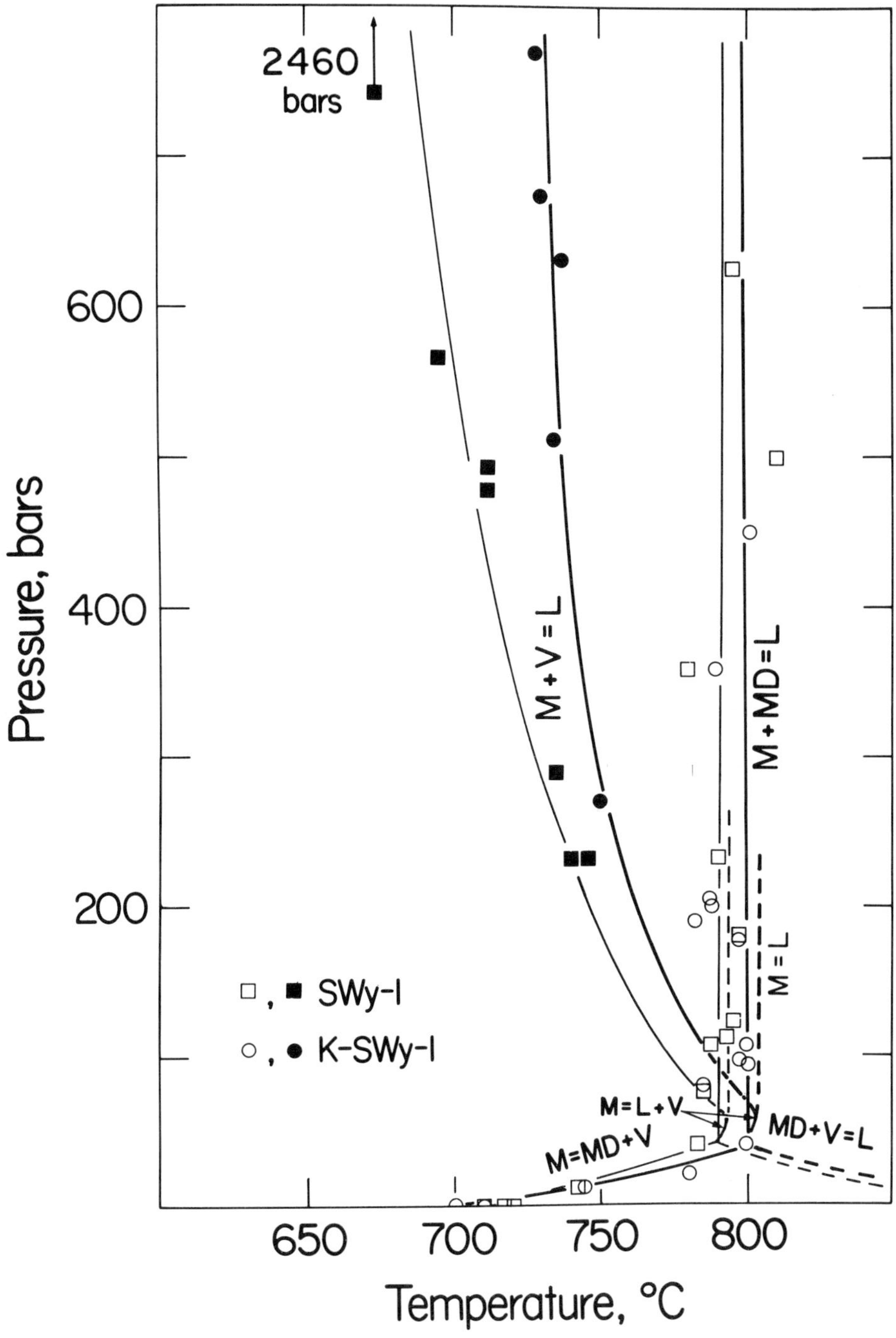

Figure 22. P-T projection of systems SWy-1-H_2O (thin lines) and K-SWy-1-H_2O (heavy lines) (Koster van Groos and Guggenheim, 1987b). M = montmorillonite; MD = montmorillonite dehydroxylate; L = liquid, V = vapor.

able for transferring gases from a 1000°C furnace at atmospheric pressure into the vacuum line of the mass spectrometer. A six-channel recorder is currently used to plot TGA, amplified (x 10) TGA, DTA, DTG, T, and a total pressure signal from a Penning gauge located at the base of the furnace (Figure 2).

Although the thermoanalyzer accommodates a sample as large as a few grams, the maximum permissible sample size for analysis in vacuum is determined by the volatile content of the sample, the heating rate, and the pumping speed of the vacuum system. In the ROM system for example, a 10-mg sample containing 5 wt. % volatiles, heated at 10°C/min, pumped with the two diffusion pumps of 135 liter/s each, typically gives very strong EGA peaks with little likelihood of vacuum overload. The TGA curve on the strip chart is readable to 5 μg and is stable to within 10 μg during analysis, which will be improved by computer control of the system.

Simultaneous DTA is possible on this instrument, but it is not normally used in high-vacuum work because the baseline is poor due to the difficulty in controlling the heating rate at temperatures <300°C. The release of adsorbed gases by a reference material during the analysis of very small samples is an additional concern.

A Mettler 1600°C furnace is used for high-temperature studies; however, analyses in high vacuum are limited to temperatures <1450°C, because at higher temperature, the pressure of O_2 and CO increases rapidly, the former presumably being produced by dissociation of alumina in the furnace lining. Crucibles of Pt-10%Rh are preferred to those of sintered alumina because they are generally easier to clean, less permeable to melts, and less adsorptive of gases at room temperature. The use of alumina crucibles may be necessary, however, for samples that can be catalyzed by Pt and for samples that react with Pt to lower its melting point or to swell or embrittle it. DTA and TGA temperature standards certified by the International Confederation for Thermal Analysis are used in thermocouple calibration to 940°C. Drop weights using pure metal wire are used for higher temperatures.

The Inficon IQ200 spectrometer utilizes a Pt-coated single-piece, ceramic mass filter and an electron multiplier. The instrument is capable of monitoring ion peak heights of 10 different masses and recording them on a strip chart. Rapid 'analog' (200-channel) scans for given mass ranges from 1 to 200 amu are also possible. The instrument operates with either constant sensitivity or constant resolution over the entire mass range.

The cold trap and diffusion pump located just beyond the sensor ensure that large quantities of evolved gases can be efficiently removed following gas analysis (Figure 2). The sensor and furnace chamber are joined by a 50-cm curved stainless steel tube. Unfortunately, this relatively long path and the large volume required to enclose the balance within the thermoanalyzer present large surface areas for the adsorption of gases. Except for one short section, the connecting tube can be baked to help remove adsorbed gases. In contrast, instruments designed to optimize EGA by mass spectrometry are free of these limitations and are capable of providing quantitative data by analyzing a direct molecular stream from the sample (Matson *et al.*, 1986). In spite of design and instrumental limitations, the quadrupole spectrometer used at the ROM provides adequate data to interpret TGA curves for most of the minerals we examine.

METHODOLOGY

Sample preparation

The investigation of various sample preparation procedures and their effects on thermal data were summarized by Mackenzie and Mitchell (1970), Smothers and Chiang (1966), and Neumann (1977). The major considerations of most methods are to remove impurities, minimize crystal structure damage due to grinding, control grain size and grain-size distribution, stabilize the sample at a specific humidity prior to analysis, select ideal sample size, maintain reproducible sample packing, match reference to sample, and evaluate the desirability of sample dilution. The first five considerations are relevant to vacuum TGA. TGA techniques at atmospheric pressures are dealt with by Bish and Duffy (1990) elsewhere in this volume.

A sample of the highest possible purity not only provides the most accurate and interpretable data, but is absolutely essential for the analysis of minor and trace volatile constituents. The effectiveness of purification procedures is governed by the initial sample purity, the amount of material available, and the physical properties of the various constituents. Standard purification methods can be employed for large samples, but the separation and manipulation of small samples require special techniques.

Modern microanalytical tools (e.g., the electron microprobe) have put new demands on older techniques such as thermal analysis. In the ROM laboratory, techniques have been developed to prepare smaller and purer samples for TGA-EGA, which will complement microprobe analyses. For coarse-grained samples, manual selection of pure mineral grains may be carried out under the optical microscope. A transparent or translucent vessel allows simultaneous transmitted and reflected illumination of the grains, making both the inclusions and coatings visible in one viewing. Slight crushing or grinding of the grains may be necessary. The sample is normally immersed in methanol (high-performance liquid chromatographic or similar low-residue grade) during the process. If the sample is only slightly impure, the impurities may be segregated and removed with a syringe.

Samples that cannot be satisfactorily purified by handpicking can be centrifuged in heavy-liquid density columns. The method has been successfully applied to samples of >1 mg to <50 mg. The effectiveness of this method decreases for grain sizes <100 μm. The sample is crushed to reduce the number of compound grains

Bailey, S. W. (1984) Structures of layer silicates: in *Crystal Structures of Clay Minerals and their X-ray Identification,* Mineralogical Society, London, 1-124.

Brindley, G. W. and Nakahira, M. (1959) The kaolinite-mullite reaction series: I. A survey of outstanding problems: *J. Amer. Ceram. Soc.* **42,** 311-314.

Cohen, L. H. and Klement, W. (1967) High-low quartz inversion: Determination to 35 kilobars: *J. Geophys. Res.* **72,** 4245-4251.

Eggler, D. H. and Rosenhauer, M. (1978) Carbon dioxide in silicate melts: II. Solubilities of CO_2 and H_2O in $CaMgSi_2O_6$ (diopside) liquids and vapors at pressures to 40 kb: *Amer. J. Sci.* **278,** 64-94.

Evans, B. W. and Guggenheim, S. (1988) Talc, pyrophyllite, and related minerals: in *Hydrous Phyllosilicates (exclusive of Micas),* S. W. Bailey, ed., *Reviews in Mineralogy* **19,** Mineralogical Society of America, Washington, D.C., 225-294.

Farmer, V. C. (1971) The characterization of adsorption bands in clays by infrared spectroscopy: *Soil Sci.* **112,** 62-68.

Grim, R. E. (1968) *Clay Mineralogy,* 2nd ed., McGraw-Hill, New York, 596 pp.

Guggenheim, S., Chang, Y.-H., and Koster van Groos, A. F. (1987) Muscovite dehydroxylation: High-temperature studies: *Amer. Mineral.* **72,** 537-550.

Harker, R. I. (1964) Differential thermal analysis in closed systems at high hydrostatic pressures: *Amer. Mineral.* **49,** 1741-1747.

Helgeson, H. G. and Kirkham, D. H. (1974) Theoretical prediction of the thermodynamic behavior of aqueous electrolytes at high pressures and temperatures: I. Summary of the thermodynamic/electrostatic properties of the solvent: *Amer. J. Sci.* **274,** 1089-1198.

Heller, L., Farmer, V. C., Mackenzie, R. C., Mitchell, B. D., and Taylor, H F. W. (1962) The dehydroxylation and rehydroxylation of trimorphic dioctahedral clay minerals: *Clay Miner. Bull.* **5,** 56-72.

Holloway, J. B. (1971) Internally heated pressure vessels: in *Research for High Pressure and Temperature,* G. C. Ulmer, ed., Springer-Verlag, New York, 217-258.

Johnson, S. L., Guggenheim, S., and Koster van Groos, A. F. (1990) Thermal stability of halloysite by high-pressure differential thermal analysis: *Clays & Clay Minerals* **38,** 477-484.

Keenan, J. H., Keyes, F. G., Hill, P. G., and Moore, J. G., eds. (1978) *Steam Tables: Thermodynamic Properties of Water Including Vapor, Liquid and Solid Phases:* Wiley, New York, 156 pp.

Khitarov, N. L. and Pugin, V. A. (1966) Behavior of montmorillonite under elevated temperatures and pressures: *Geokhimiya* **7,** 790-795 (transl.), *Geochem. Int.* **3,** 621-626.

Koster van Groos, A. F. (1979) Differential thermal analysis of the system NaF-Na_2CO_3 to 10 kbar: *J. Phys. Chem.* **83,** 2976- 2978.

Koster van Groos, A. F. and Guggenheim, S. (1984) The effect of pressure on the dehydration reaction of interlayer water in Na montmorillonite (SWy-1): *Amer. Mineral.* **69,** 872-879.

Koster van Groos, A. F. and Guggenheim, S. (1986) Dehydration of K-exchanged montmorillonite at elevated temperatures and pressures: *Clays & Clay Minerals* **34,** 281-286.

Koster van Groos, A. F and Guggenheim, S. (1987a) Dehydration of a Ca- and a Mg-exchanged montmorillonite (SWy-1) at elevated pressures: *Amer. Mineral* **72,** 292-298.

Koster van Groos, A. F. and Guggenheim, S. (1987b) High-pressure differential thermal analysis (HP-DTA) of the dehydroxylation of sodium- and K-exchanged montmorillonite: *Amer. Mineral.* **72,** 1170-1175.

Koster van Groos, A. F. and Guggenheim, S. (1989) Dehydroxylation of Ca- and Mg-exchanged montmorillonite: *Amer. Mineral.* **74,** 627-636.

Koster van Groos, A. F. and ter Heege, J. P. (1973) The high-low quartz transition to 1 kilobar pressure: *J. Geol.* **81,** 717-724.

Le Chatelier, H. (1887) De l'action de la chaleur sur des argile: *Bull. Soc. Franc. Mineral. Cristallogr.* **10,** 204-210.

Low, P. F. (1979) Nature and properties of water in montmorillonite water systems: *Soil Sci. Soc. Amer. J.* **43,** 651-658.

The details of coupling a mass spectrometer to a thermoanalyzer were described by Clinckemaille and Hofmann (1972), Szekely and Till (1974), and Dunner and Eppler (1975). Wendtlandt (1986) recently reviewed thermobalance EGA instrumentation and TGA-EGA techniques and applications. Dollimore *et al.* (1984), Holdiness (1984), and Kaiserberger and Emmerich (1982) reviewed and specifically discussed EGA by mass spectrometry, and Morgan *et al.* (1988) and Gallagher (1982) reviewed applications of EGA to earth science and inorganic material studies, respectively. Other relevant papers on high-temperature topics, such as MS analysis of molecular beams, effusion cells, and high-pressure sampling techniques using skimmers, can be found in Hastie (1979).

INSTRUMENTATION

Thermal balance

The thermal balance is the central component of a TGA-EGA system. Each thermal balance on the market has its own particular configuration with specific advantages and disadvantages. The choice of instrument depends on the application and, if necessary, on how the balance must be modified to accommodate the desired EGA instrumentation. Many instruments are compactly constructed and are designed specifically for atmospheric TGA; few are designed for high-vacuum studies. If neither quantitative EGA nor measurement of readily condensable volatiles, such as alkalies, halides, or various metals, are of interest, the gas analyzer need not be in line-of-sight with, nor in close proximity to the sample, and modification may be fairly simple. More demanding applications, such as dynamic, high-temperature analysis of trace volatiles in a milligram-size sample, require an instrument designed with sub-microgram sensitivity, an ultimate vacuum better than 10^{-7} torr, a balance mechanism unaffected by pressure changes, a short line-of-sight path from the sample surface to the detector, a furnace chamber of small volume, and good heating control to 1400°C. In practice, major difficulties, such as the alignment of the detector with the sample, are likely to be encountered in the modification of an off-the-shelf instrument. If, in addition, the same instrument also must be capable of operating at atmospheric pressure and performing differential thermal analysis (DTA), further problems will arise. The difficulty in modifying an instrument increases with the addition of increased capabilities, and some problems may be best solved by the use of more than one instrument. For example, a conventional vacuum TGA-DTA instrument might be used in conjunction with a Knudson cell-mass spectrometer system.

Mass spectrometer

Several gas analysis techniques are available for thermal analysis: mass spectrometry (MS), Fourier-transform infrared (FTIR) spectrometry and other infrared methods (Low, 1967), gas chromatography (GC), manometry (Harris, 1981), and an assortment of specific gas detectors. GC and FTIR spectrometry are increasingly used alone and in combination with MS. Prime (1987) has compared the merits and limitations of MS and IR in conjunction with TGA; however, for vacuum analysis the choice is practically restricted to MS.

Several mass spectrometers are available to match specific analytical problems. The dynamic nature of TGA-EGA requires an analyzer capable of rapid gas analysis with moderate precision, sensitivity, and stability. The quadrupole mass spectrometer with its rapid response best meets these requirements. Some analyses, involving high mass numbers or very complex decompositions, may require the greater resolution provided by magnetic-sector mass spectrometers. These are more expensive and generally slower, thus sacrificing rapid response. Some single-focusing magnetic mass spectrometers are capable of resolving ion doublets and triplets, such as mass 14 (N, CH_2), 16 (O, NH_2, CH_4), 28 (CO, N_2), and 32 (O_2, S), which are most likely to be encountered in mineral analyses (Elliot, 1969). This resolution is beyond the capabilities of quadrupole mass spectrometers, although the problem can be addressed, but with less precision, by the analysis of cracking patterns (discussed below).

The need for high sensitivity can be best met by placing the spectrometer ion source in line with the specimen surface, using a suitable beam skimmer, and using an ion source designed specifically for molecular beam analysis. The low-end magnetic-sector mass spectrometers are cost competitive with the moderate-end to high-end quadrupole instruments (Rosen *et al.*, 1988), but quadrupole mass spectrometers provide a good balance of fast response to the dynamic heating of the system, a reasonable analytical range of masses, and good sensitivity at a reasonable cost.

The theory of quadrupole mass spectrometry was discussed by Dawson (1976). Basic quadrupole spectrometers operate by bombarding a gas at low pressure (generally $<10^{-4}$ torr) with electrons of sufficient energy to allow molecular ionization. In the process, a certain proportion of the molecules is broken into ionized fragments. Positively charged fragments are usually used, but some instruments can also use negatively charged fragments. The charged molecules and fragments are accelerated and then passed through a filter of radio frequency (RF) and direct current (DC) fields toward a detector. For a given RF-DC combination, only ions of a specific mass/charge (m/e) ratio will successfully oscillate through the filter to the detector. The instrument rapidly steps through all necessary combinations to produce a mass spectrum over a given range of m/e. The mass values and peak intensities for the parent molecule and all fragments of the parent molecule are referred to as the cracking pattern for the gas. The mass spectrum is the sum of cracking patterns of all gases present. Some caution must be used in interpreting mass spectra because, as shown by Breth *et al.* (1983), cracking patterns of mixtures vary with pressure.

THERMOGRAVIMETRIC ANALYSIS OF MINERALS

DAVID L. BISH[1] AND CLARENCE J. DUFFY[2]

[1]Earth and Environmental Sciences Division, Mail Stop D-469
[2]Isotope and Nuclear Chemistry Division, Mail Stop J-514
Los Alamos National Laboratory
P.O. Box 1663
Los Alamos, New Mexico 87545

INTRODUCTION

Thermogravimetric analysis (TGA), in which the mass of a sample is monitored as a function of temperature, is one of the oldest analytical techniques used in clay mineralogy. One of the first applications to the study of minerals was reported in 1903 by Nernst and Riesenfeld. Since then, TGA has been used to obtain a variety of information on minerals, particularly hydrous phases, such as clays and zeolites. TGA is currently widely used in many other disciplines as well, including polymer chemistry, pharmaceuticals, and inorganic analysis (Wendlandt, 1986). Despite the active interest in TGA in other areas of solid characterization and despite the many studies of clay minerals that have been conducted using TGA, the method has not enjoyed significant popularity in recent years in the mineral sciences. The technique appears to be viewed as somewhat qualitative in a field that is demanding more and more quantitative results. Fortunately, however, as a result of the large amount of research in other fields employing TGA, the application of TGA has grown in recent years from a simple technique often used as a fingerprint or water-analysis method to one that can provide quantitative information on many types of thermal processes in minerals. Data available through TGA include, but are not limited to, the kinetics of dehydration and dehydroxylation reactions, the analysis of solids for non-water volatiles, the determination of equilibrium dehydration behavior of hydrous minerals, the quantitative analysis of multicomponent mixtures, and the separation of overlapping dehydration reactions.

EXPERIMENTAL

Thermal data presented in this chapter were obtained on a variety of natural and synthetic samples. Sources of natural materials included the Source Clays Repository of The Clay Minerals Society; Ward's Natural Science Establishment, Rochester, New York; Minerals Research, Clarkson, New York; and Harvard University Mineral Museum (samples labeled HU), as well as samples collected by the authors. All materials were characterized for purity using quantitative X-ray powder diffraction (XRD) analyses before thermal analysis experiments were performed. XRD used an automated Siemens D-500 diffractometer. Samples were pure unless otherwise stated.

Thermal analyses employed a DuPont Model 1090 system equipped with a Model 951 thermogravimetric analyzer. Heating rates, sample sizes, and gas flow rates for individual experiments are given with the data. Derivative TGA curves were calculated digitally from the raw TGA data and were reproducible for different samples of the same material. Calibration of the thermogravimetric analyzer was performed using independently calibrated weights and thermocouples traceable to the National Institute for Standards and Technology.

FUNDAMENTALS OF THERMOGRAVIMETRIC ANALYSIS

A variety of TGA equipment is commercially available today that commonly utilizes computer-interfaced high-precision analytical balances coupled with solid-state temperature controllers. Experimental conditions can therefore be closely controlled and monitored, and precise data can be obtained on very small samples. Computer control of data acquisition allows easy storage of data for later analysis by numerous techniques. The variety of TGA equipment on the market today was discussed by Wendlandt (1986). With appropriate modifications, equipment currently available can be utilized to study thermal reactions of minerals under a wide range of conditions, from isothermal to heating rates as great as several hundred degrees per minute, from high vacuum to elevated pressures (see Wendlandt, 1986, 130-131), and in a variety of atmospheres.

Characteristics of mass-loss curves

Data resulting from dynamic (nonisothermal) thermogravimetric analysis, in which the temperature is varied at a constant rate, typically exhibit several common features. Wendlandt (1986) described two temperatures characteristic of all single-stage, nonisothermal reactions: T_i, the initial temperature (or procedural decomposition temperature), and T_f, the final temperature. These temperatures are shown in the TGA curve (Figure 1a) of $Na_2WO_4{\cdot}2H_2O$ obtained at a heating rate of 10°C/min with an N_2 flow rate of 50 cm^3/min. T_i is the temperature at which the mass change is first detectable, and T_f is the temperature at which the mass change is complete.

Neither T_i nor T_f has any physical significance, but several characteristics may be noted about these two temperatures. T_f is greater than T_i, and T_f - T_i is called the reaction interval. For an endothermic reaction, T_i, T_f, and the reaction interval increase (or stay the same) with increasing heating rate. Figure lb, also of $Na_2WO_4{\cdot}2H_2O$, illustrates additional temperatures that are also characteristic of a nonisothermal reaction, namely T_o, T_{in}, and T_e, the extrapolated onset temperature,

455° and 555°C. Brydon and Turner (1972) used a modest vacuum of 0.5 mm Hg (0.5 torr) for their DTA run, so a direct comparison of the two sets of results is difficult. The multiple midrange peaks in Figure 3 either could indicate the presence of other phyllosilicates, but because Brydon and Turner found none, the total pressure curve probably instead illustrates the complex, multi-stage nature of the dehydroxylation reaction. An increased sensitivity to fine details, such as these, is typical of high-vacuum TGA-EGA studies.

In the high-temperature region (600°-900°C), the shape of the total pressure curve indicates that dehydroxylation continues above 600°C, and water-loss peaks can be seen at 807° and 848°C in Figure 3. The 807°C peak corresponds with the 822°C endotherm recorded by Brydon and Turner (Figure 4), but it is in part produced by a minor loss of CO_2 (not shown), possibly from a minor impurity. The 848°C peak is related to the final loss of hydroxyl during recrystallization and corresponds well with the 856°C exotherm recorded by Brydon and Turner.

Should the division of a complex weight loss from a single volatile species into steps be necessary, points of minimal partial pressure may be used in the same manner as, but more precisely than, DTG minima. The division of the losses at a total pressure minimum or a DTG minimum would probably result in an erroneously high value for the first volatile and a correspondingly low value for the second because weight losses typically commence more gradually than they terminate. Thus, the rapidly decreasing loss due to one volatile compound may overlap the gradually increasing loss of a second. Because such asymmetry is evident in most evolved gas curves, a more accurate temperature can be estimated from the point at which the tail area of the first volatile is equivalent to the onset area of the second volatile, with consideration being given to relative ionization probabilities, yields, and masses.

If losses of two volatiles are strongly overlapping or simultaneous, only the combined weight-loss percentage normally can be expressed, although a ratio may be approximated from peak areas on evolved gas curves. For minor partial pressure peaks, the proportion of that volatile can be estimated to within an order of magnitude. If very minor peaks are of interest, their source needs to be determined because they may: (1) originate from the sample, (2) be an artifact of adsorbed gases being stripped from system surfaces by molecules of the principal volatile, or (3) have formed by reaction among evolved gases, residual gases, and system surfaces. Comparative analyses of appropriate characterized materials will likely be necessary to solve this type of problem.

The majority of hydrated minerals evolve H solely as H_2O (Figure 3); however, a significant number of minerals have been found to evolve H_2 and HF in addition to H_2O, but rarely hydrocarbons or HCl. The presence of other H-bearing species may seriously complicate an otherwise straightforward analysis for water. The H_2 loss at 360°C from maricopaite, $Pb_{7.00}Ca_{2.17}(Si_{34.46}Al_{11.34})O_{97.10}\cdot 30.96H_2O$, (Peacor *et al.*, 1988) was unrelated to the H_2O loss at 100°C or the water loss at

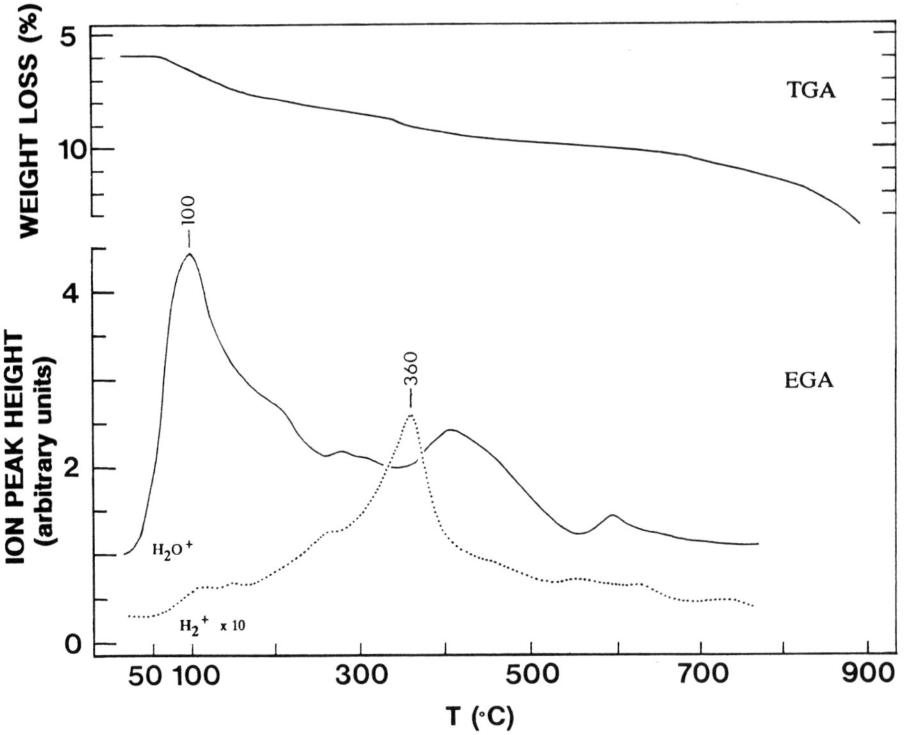

Figure 5. Thermal gravimetric-evolved gas analysis curves of maricopaite, Tonopah, Arizona, (2.54 mg), heated at 10°C/min, initial vacuum 2 x 10^{-8} torr.

410°C (Figure 5). The source of the H_2 was unknown. This problem was complicated in that most of the available material (2.54 mg) was used for the first thermogram. Fortunately, an additional small amount of sample was made available; a pure 0.565-mg sample was obtained by density gradient separation in a centrifuge. The repeat analysis indicated that the source of the H_2 was hydrocarbons presumably occluded within the zeolite-like cages in the structure.

Hydrogen is commonly lost from hydrous Fe-bearing minerals. The thermogram (Figure 6) of the unusual ferroan talc, described by Robinson and Chamberlain (1984), $(Mg_{1.90}Fe^{2+}{}_{0.78}Fe^{3+}{}_{0.26}Al_{0.02}\,\square_{0.04})Si_{4.04}O_{10.06}(OH)_{2.23}$, illustrates this phenomenon. Hydrogen begins to evolve at 500°C and peaks at 680°C. Water produced by the dehydroxylation of the talc begins to evolve at 600°C and peaks at 755°C. The second small H_2^+ peak at 755°C on the main H_2^+ peak is a cracking fragment of the H_2O which produces the H_2O^+ peak of 755°C. The evolution of H_2 appears related to a local auto-oxidation-dehydroxylation reaction involving Fe^{2+} and OH. Dehydroxylation is followed by an O_2 loss over 1000°C due to the re-duction of Fe^{3+} (shown only on the TGA curve on Figure 6). The dehydroxylation peak temperature in talc is inversely proportional to iron content: 755°C for the ferroan talc (16.8 wt. % total Fe as FeO, Figure 6), 845°C for moderate-iron talc (2.3 wt. %

One of the most important applications of DTG to studies of clays and related materials is the separation of overlapping reactions. If several mass-loss reactions occur over a narrow temperature range, they yield TGA curves with no resolved plateaus. The TGA curves of the zeolite clinoptilolite (Figures 17 and 18) exemplify this effect well. The use of DTG curves allows one to distinguish between overlapping reactions that otherwise might be difficult to discern using TGA data alone. These reactions may occur at different rates, yielding DTG curves of different peak heights. In practice, shapes and temperatures of DTG curves commonly resemble those of differential thermal analysis (DTA) curves or differential scanning calorimetry (DSC) curves of the same material. DTA and DSC peaks, however, are typically displaced to higher temperatures than the corresponding DTG peaks, because of the significant differences in the atmospheres in TGA, DTA, and DSC instruments. DTA and DSC experiments are often performed in sealed or packed containers, which allow a significant build-up of evolved gas pressure, whereas TGA experiments are usually conducted on loose powders in open pans, allowing little pressure build-up. Comparisons of TGA and DTG curves for overlapping reactions were presented by Wendlandt (1986, p. 55).

Sources of mass loss or gain

Because TGA measures the mass of a sample, any volatile species that are gained or lost during analysis will be measurable. In the mineral sciences, these species are usually limited to gaseous species, such as CO_2, CO, O_2, H_2O, NH_3, and SO_2. A variety of other species, such as $(CN)_2$, HCl, CH_3COOH, or even metal ions, may also be lost if unusual inorganic or organic compounds are analyzed. Many reactions seen with TGA can be envisioned as the loss of molecular species from the crystal. Gypsum ($CaSO_4 \cdot 2H_2O$), for example, undergoes a reaction above 90°C to anhydrite ($CaSO_4$), with the evolution of gaseous H_2O (Figure 2). Although the water molecules in gypsum are held in crystallographically well-defined sites, the situation is similar to that found for the low-temperature reactions of smectites and zeolites, in which molecular water, held with a variety of energies in interlayer or cavity sites, is released upon heating.

The solids of interest, however, need not contain the simple species listed above, and the particular gases that evolve upon heating commonly result from reactions within the sample, i.e., they are not present as discrete species before heating. The thermal reactions of Ca-oxalate monohydrate ($CaC_2O_4 \cdot H_2O$) are a good example of this. The TGA curve in Figure 3 shows three major mass-loss events. The first (DTG max at ≃170°C) corresponds to the dehydration of Ca-oxalate monohydrate,

$$CaC_2O_4 \cdot H_2O(s) \rightleftharpoons CaC_2O_4(s) + H_2O(g)\uparrow . \tag{1}$$

The second mass loss (DTG max at ≃476°C) results from a reaction of dehydrated Ca-oxalate,

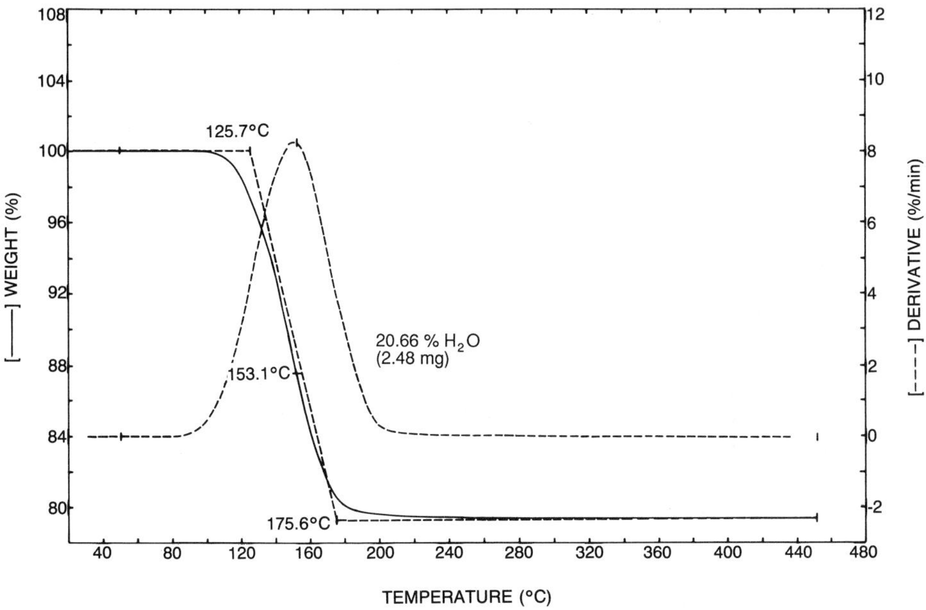

Figure 2. Thermogravimetric analysis curve of natural gypsum showing mass loss due to evolution of H_2O between 100° and 220°C (11.99-mg sample, 200-mesh particle size, 20°C/min heating rate, 50-cm^3/min N_2 purge).

$$CaC_2O_4(s) \rightleftharpoons CaCO_3(s) + CO(g)\uparrow . \quad (2)$$

The third mass loss in Figure 3 (DTG max at $\simeq$703°C) results from the breakdown of $CaCO_3$,

$$CaCO_3(s) \rightleftharpoons CaO(s) + CO_2(g)\uparrow . \quad (3)$$

Clearly, Ca-oxalate does not contain the simple species CO and CO_2 in its structure; these gases are formed during the thermal reaction.

The thermal reactions of Ca-oxalate monohydrate also illustrate another phenomenon, that of reaction between evolved gases and the purge gas in the thermobalance. Examination of the gases evolved during this reaction (e.g., with a mass spectrometer) reveals that O_2 present in the thermobalance will react with the CO evolved during the second thermal reaction. The exothermic reaction of O_2 with CO will raise the sample temperature and thus accelerate the breakdown of Ca-oxalate.

Another example of the production of species different from those contained in a sample is the reaction that takes place on heating kaolinite ($Al_2Si_2O_5(OH)_4$)

Vacuum thermogravimetric analysis

The furnace chamber, the spectrometer sensor, the vacuum connection between them, and the crucible (Figure 2) are baked in high vacuum just before the analysis to minimize contamination, especially any 'memory' of volatiles from the previous analysis. The sample is weighed and immediately inserted into the furnace chamber. Contaminating air is then purged from the vacuum system with high-purity N_2 (<2 ppm each of O_2 and H_2O). The system is pumped until the vacuum is better than 1 x 10^{-7} torr. This takes 1-4 hr, depending upon the quantity of loosely held volatiles evolving from the sample. The difference between zero and the position of the beginning of the TGA curve on the ordinate of the thermograms (Figures 1, 3, 5-7, and 9) represents the loss incurred during the initial pumpdown to vacuum at room temperature. Once a steady vacuum is produced, the sample is heated at 10°C/min, but the heating rate may be decreased or increased if the quantity of volatiles is anticipated to be very large or very small. The risks of overwhelming the high vacuum system with an overly rapid volatile loss and of sample ejection caused by rapid volatile loss must be kept in mind. The analysis is normally allowed to continue until no further volatiles evolve or until a major weight loss occurs with no evident increase in any partial pressure, which would indicate that the evolved volatile is condensing on the chamber walls and not reaching the sensor of the spectrometer.

At the end of an analytical run, the sample is cooled, the system is repressurized with N_2, and the sample is quickly reweighed. The total loss determined by TGA is subtracted from the total loss determined by microbalance. The difference represents the weight of loosely bound volatiles lost at room temperature in vacuum prior to analysis. Inasmuch as this value is calculated using three weight measurements, the cumulative error is a relevant factor for samples weighing <5 mg. The method is satisfactory only if the thermal product does not absorb appreciable N_2 or H_2O upon repressurization. For zeolites, heating the sample to temperatures >800°C to destroy the framework structure is advisable.

Because evolving gases are uninhibited by atmospheric pressure, weight losses are accelerated in vacuum. For example, a CO_2 loss by the rare-earth silicate tundrite, $Na_3(Ce,La)_4(Ti,Nb)_2(SiO_4)_2(CO_3)_3O_4(OH)\cdot 2H_2O$ (Karup Moller, 1982), is quite rapid, even at atmospheric pressure; in vacuum this mineral loses 15 wt. % over a 25°C interval at a heating rate of 10°C/min! To avoid sample ejection during such losses, even small samples are normally run in tall crucibles with cocked lids. Crucible walls and lids are checked at the end of a run to determine whether any of the sample may have been ejected. Although any loss of sample would ruin a weight-loss measurement, the temperatures of the weight-loss events and curve shape may still be useful.

If a thermal product is to be identified, the sample can be annealed at the critical temperature for several minutes at the end of the heating run, cooled in vacuum, weighed, and immediately analyzed by XRD. Cooling rates in vacuum are relatively slow; hence, samples may quench poorly. Glass capillaries can be used to encase humidity-sensitive products for analysis by XRD.

Evolved gas analysis

The quadrupole mass spectrometer used in the ROM system can be operated in two modes: continuous scanning over any range between mass 1 and 200 and continuous monitoring of as many as 10 pre-programmed channels. All components of the instrument that can be baked are maintained at 100°C during analysis to minimize adsorption of evolving gases on the exposed surfaces. In mineral analysis six masses, namely, 2 (H_2^+), 18 (H_2O^+), 28 (N_2^+, CO^+), 32 (O_2^+), 44 (CO_2^+), and 64 (SO_2^+), are routinely monitored along with any other anticipated masses. Analysis proceeds with periodic scans over the full mass range, particularly near the onset and the peak of a weight loss, to determine if any new volatile species are evolving. A characteristic mass for each new volatile is selected, entered on a channel, and continuously monitored. On the present ROM equipment this procedure is done manually, which causes a break in data collection of 1-2 min while a scan is collected and evaluated. Under software control, the spectrometer may resume routine monitoring immediately following the full spectrum scan, while the software performs the full spectrum analysis and selects the additional masses to be monitored.

The peak corresponding to the parent molecule is usually monitored, unless a characteristic fragment is much more abundant. For example, SO_3 produces a very weak peak relative to the SO_2^+ fragment due to the high degree of dissociation under these analytical conditions (Langer and Gohlke, 1967). Here, the SO_2^+ fragment is monitored.

A further consideration is that a peak for a given mass may not be unique to a specific volatile. For example, mass 32 may be O_2^+, $^{32}S^+$, or a mixture of the two. To resolve this, the mass 34/32 ratio must be determined. This ratio is 0.002 for O_2 and 0.042 for S. If the ratio indicates the presence of S, the mass spectrum is examined for the characteristic peaks of H_2S, SO_3 (inferred by the presence of SO_2), or S_8. Incidentally, a sulfur vapor spectrum, such as given by Langer and Gohlke (1967), might be misinterpreted as an SO_2 spectrum should it be recorded by a mass spectrometer having a low mass range (1-50 amu).

Instruments having components that cannot be baked (e.g., the balance chamber in the ROM instrument) will likely have H_2O as the most abundant residual gas, followed by N_2, O_2, CO_2, and H_2. As a consequence of the high H_2O background, relatively large samples (10-50 mg) must be analyzed to detect trace or minor water (0.X - 0.0X wt. %). Furthermore, the cracking pattern for water shows abundant fragments at masses 16 (O^+) and 17 (OH^+). These masses interfere with the peaks for methane and ammonia parent molecules, respectively, so the less-abundant fragments CH^+ and NH^+ must be monitored if these gases are present. The presence of residual hydrocarbons from vacuum pumps and seals typically con-

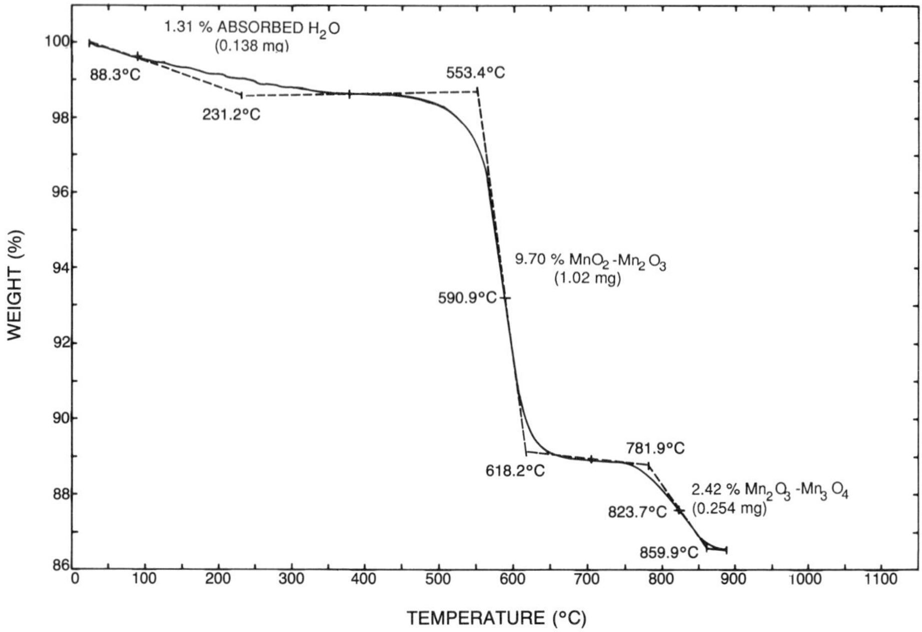

Figure 5. Thermogravimetric analysis curve of pyrolusite (MnO_2) (Lake Valley, New Mexico, HU 111929), showing mass losses due to evolution of adsorbed water and O_2 as a result of transformation to Mn_2O_3 and Mn_3O_4 (17.67-mg sample, 10°C/min heating rate, 50-cm³/min N_2 purge). Calculated mass loss for the first reaction is 9.2%, for the second reaction, 3.1%. Lower observed value for the second reaction is likely due to incomplete reaction to Mn_3O_4 at 900°C.

$$CaSO_4 \cdot 2H_2O \rightleftharpoons CaSO_4 + 2\,H_2O(g)\uparrow . \tag{6}$$

One can write an equilibrium constant for the reaction, as follows:

$$K = \frac{[CaSO_4][H_2O]^2}{[CaSO_4 \cdot 2H_2O]} , \tag{7}$$

where the brackets around the reactants and products denote the activity of the species. Using an activity of 1.0 for the solid reactants and products, Eq. (7) reduces to $K = [H_2O]^2$. The equilibrium constant is therefore a function only of the square of the H_2O vapor pressure (or fugacity) and, thus, it is easy to understand that at equilibrium, increasing the water vapor pressure within the sample will drive the reaction to the left, i.e., will inhibit the reaction and cause reaction temperatures to be higher. A variety of factors affect the water vapor pressure within the sample, as will be discussed below.

The other important aspect of thermal reactions is the kinetics of processes occurring in the thermobalance. These ongoing processes can include at least three reactions: (I) destruction of bonds involving the volatile component, (II) diffusion of this released volatile through the solid, and (III) transport of the volatiles from the pore spaces in the powder to the purge gas in the thermobalance. These processes vary in relative importance for different conditions and sample geometries, and they can be expanded to include many other steps. The third reaction, for example, can include transport by Knudson diffusion, Soret diffusion, or convection. Additional complexity is introduced if temperature is varied during the TGA experiment, because reactions I-III are thermally activated with different activation energies.

If the experiment can be designed such that two of the above reactions are fast relative to the third, the kinetics of this slowest step can be examined directly using TGA. The interpretation of typical results, however, requires that the processes of volatile transport in reactions II and III be understood. For example, the treatment of data would be considerably different if volatiles were convected through the sample powder (Rosen, 1952) than if transport were predominantly by diffusion (Crank, 1975; Calef and Deutsch, 1983). In the latter situation, the rate of volatile evolution from the sample powder depends on several factors, including the porosity of the powder, the shapes of the crystals, and the diffusion coefficients for volatiles in the crystals and through the powder.

Most of these processes depend explicitly on volatile concentration. For samples in which rates of volatile loss are controlled by reaction I above, *steady state* rate laws are commonly observed (Lasaga, 1986). For these rate laws, both the Gibbs free energy and the rate coefficient depend explicitly on temperature. Thus, *steady state* rate laws require isothermal conditions. The Gibbs free energy can be related to the mass action relation for the experimental conditions and for the standard state. The most important variable in this mass action equation, as shown in Eq. (7), is the fugacity of the volatile component relative to some standard state. The rate of the devolatilization reaction thus depends on the fugacity of the volatile immediately adjacent to the mineral. Transport of volatiles is also affected by the fugacity adjacent to the mineral, even if reaction I is rapid. Thus, reactions I-III above potentially exhibit a dependence on the fugacity of the volatile adjacent to the mineral.

Traditionally, the basis for obtaining kinetic results from dynamic TGA data is the well-known kinetic equation (Sestak, 1966),

$$-dX/dt = kX^n, \tag{8}$$

where X is some unit of concentration or amount, t is time, k is the rate constant, and n is some real number, often called the order of the reaction. Different reactions or rate laws will apply to each of the reactions occurring in the individual sample particles or crystallites and in the polycrystalline mass of powder (reactions I-III). It is very important to note that the several reactions occurring in a TGA sample are usually coupled and interdependent and that the slowest reaction will limit the overall thermal reaction. The intrinsic thermal reaction being observed (dehydration, break-thermal conductivity in vacuum. The endotherms involving loss of volatiles are further weakened by the efficient removal of the evolving gas by the vacuum pumps.

SUMMARY

Advantages of the vacuum method

1. If a thermoanalyzer is operated in vacuum, the TGA baseline is flat and the `buoyancy' effect present during analysis at atmospheric pressure is eliminated. The buoyancy effect is caused by the convection and the reduction in density of the gas around the crucible as it is heated, which produces an apparent weight gain. Although compensation for the buoyancy effect is possible by subtraction of a TGA curve for an equivalent volume of inert material from the TGA curve of the sample analyzed, the error may be large at lower temperatures where the effect is most pronounced. Most modern microbalance designs have reduced or eliminated this problem.

2. At pressures $<10^{-4}$ torr, diffusion from sample surfaces is effectively uninhibited, and volatiles are efficiently removed from the sample during decomposition. This efficient removal of volatiles reduces the influence of sample size on decomposition temperatures and improves the reproducibility and comparability of successive analyses.

3. Evolved gases are easily and efficiently directed from the furnace to the gas analyzer without dilution so as to produce a relatively high sensitivity and instantaneous response for EGA. The problems introduced by impurities in a carrier gas are eliminated.

4. The oxidation of a sample can be controlled in vacuum. Ferrous compounds tend to oxidize even at moderate vacuum (10^{-3} torr), causing gradual weight gains at moderate temperatures (400°-600°C). The O_2 causing this may originate from air leaks in the system, incomplete flushing of air from the sample, or minor O_2 impurities in a flushing gas. At pressures $<10^{-6}$ torr, however, oxidation does not occur in minerals that are otherwise prone to oxidation. (An atmosphere containing 5% H_2 would have a similar effect, but analysis for H would not be possible.)

5. The use of conventional H_2O- and H_2O+ terms to express the state of water in an analysis may serve a purpose in rock analysis, but it may well be misleading or uninterpretable when applied to mineral analysis. An expression of water content in terms of H_2O lost and H_2O retained in vacuum is generally more interpretable and is particularly of value in interpreting electron microprobe analyses of hydrated minerals. It would be easy to design a simple vacuum system for determining H_2O lost and H_2O retained in vacuum for whole rock analysis.

6. Sorption-desorption and reactivity of vacuum-heated solids can be conveniently studied. A sample can be heated in vacuum to a specific temperature, a gas can be admitted into the vacuum system for a given period, the sample can be cooled, and TGA-EGA can then be run, all on the same instrument.

7. Operation in vacuum reduces the importance of grain size and packing so that comparative analyses are much easier to make. For example, calcite thermograms in vacuum typically shows decomposition peaks in a narrow range of 650°-665°C, whereas in atmosphere the range is considerably greater (860°-1010°C; Webb and Heysteck, 1957) due to the effect of differences in grain size and packing.

8. The recorded total pressure and partial pressure curves of evolved gases in vacuum analysis are less noisy and, hence, more sensitive than similar curves recorded from analysis in atmosphere or from other techniques, such as DTG.

9. Detection of H_2, F_2, and other reactive volatiles may be possible only in vacuum studies. These volatiles generally react with other gases or the walls of the instrument during analysis at atmospheric pressure prior to transfer to the gas analyzer.

Disadvantages of the vacuum method

1. The weight loss of volatiles from a sample as the vacuum system is being pumped down cannot be directly measured, and the species causing the loss cannot be identified with the mass spectrometer. Therefore, the detailed study of intermediate hydrates involving loosely bound volatiles requires analysis at atmospheric pressures.

2. Major fluctuations in evolved gas pressure during decomposition will affect the TGA baseline of some balances in vacuum. For many materials, this may not be a problem because critical measuring points in most weight-loss determinations lie before, between, and after weight-loss events and, thus, are on the baseline and are unaffected by pressure changes. A weight value at the dividing point between poorly resolved losses, however, will be affected. Normally, a correction can be applied to this value based on the pressure of the volatile at that point (see Oswald and Wiedemann, 1977, p. 159). If accurate weight values are required throughout the analysis, as in studies of kinetics, the balance must be of a design unaffected by pressure changes. If this is not possible, the effect might be minimized by the use of a constant leak (He, Ar, or N_2), so that the residual gas pressure would be about an order of magnitude greater than any evolved gas pressures. Alternatively, a slower heating rate might help solve or minimize the problem.

3. Weight losses occur at lower temperatures in vacuum particularly in the higher-temperature range; thus, much more overlap of weight losses may exist than in

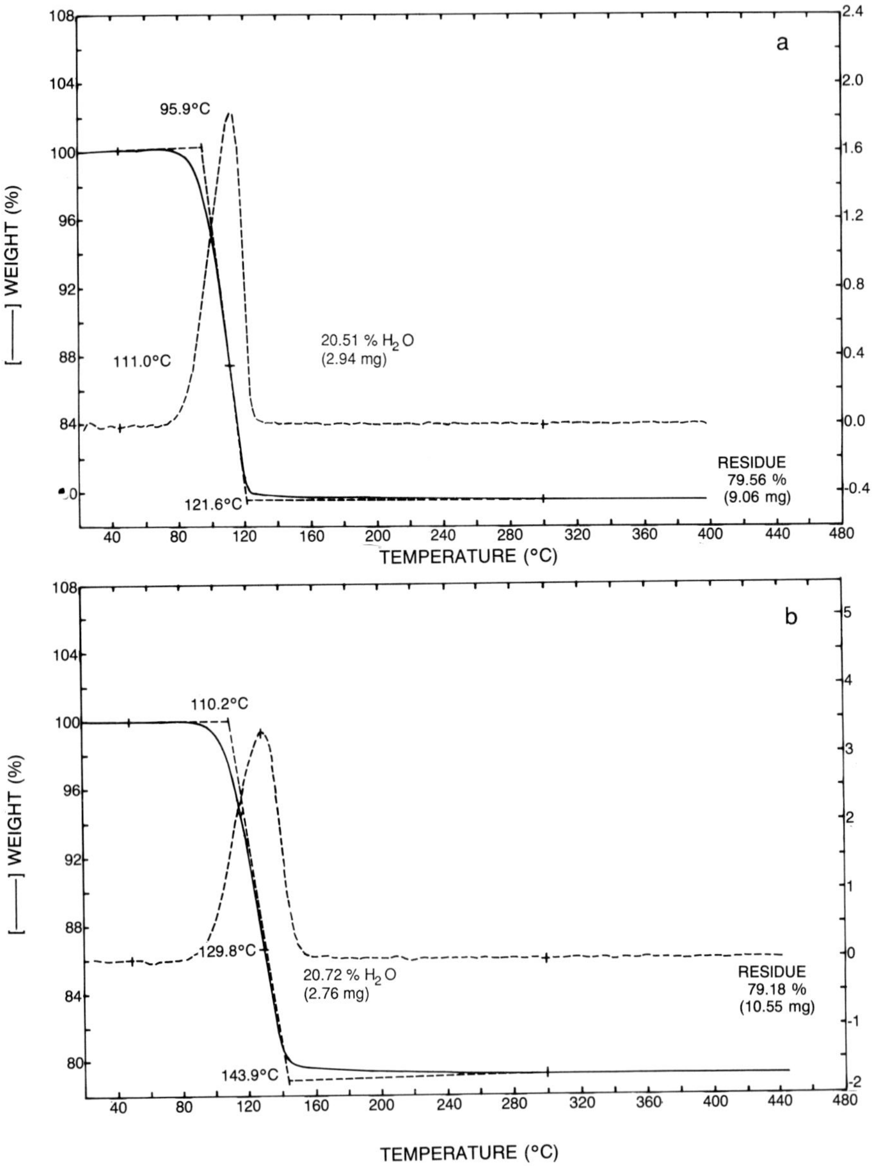

Figure 6a, b. Thermogravimetric analysis curves of natural gypsum collected at heating rates of (a) 2°C/min (11.39-mg sample) and (b) 5°C/min (13.33-mg sample), showing increase in dehydration temperature with increase in heating rate. Both samples heated in a 50-cm^3/min N_2 purge.

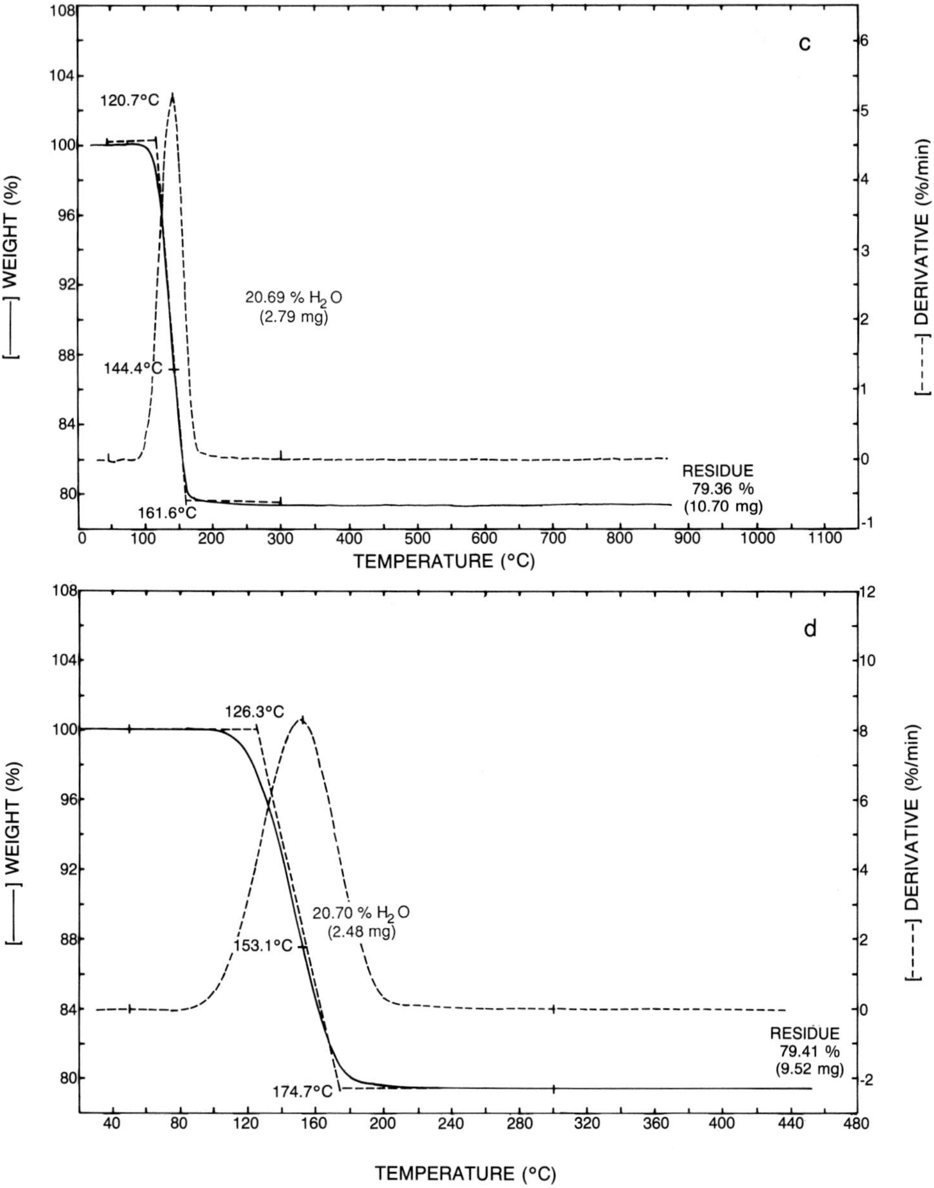

Figure 6c, d. Thermogravimetric analysis curves of natural gypsum collected at heating rates of (c) 10°C/min (13.49-mg sample) and (d) 20°C/min (11.99-mg sample), showing increase in dehydration temperature with increase in heating rate. Both samples heated in a 50-cm^3/min N_2 purge.

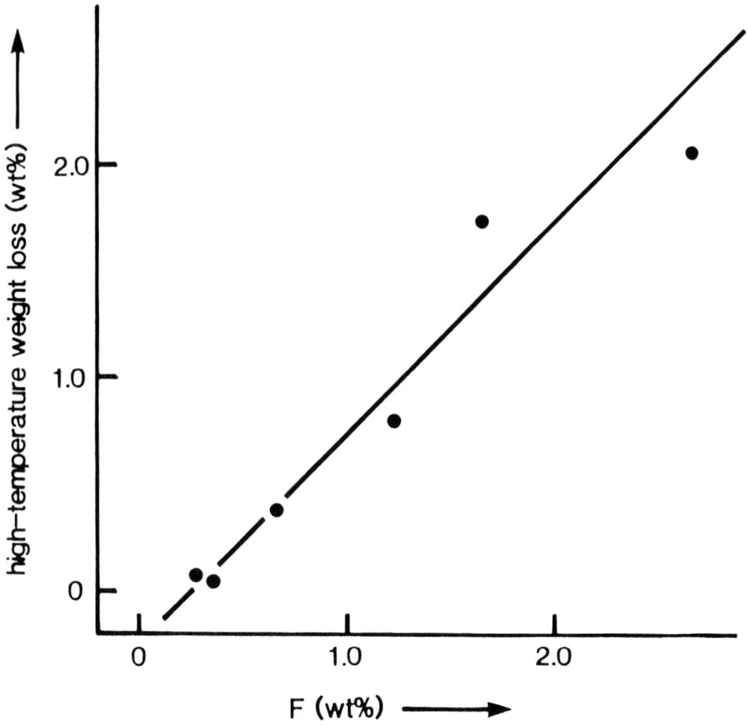

Figure 8. Plot of high temperature (>1100°C) weight loss (wt. %) vs. F content (wt. %) of titanite specimens from various localities. From Hawthorne *et al.* (1990).

was discussed by Nikolaev *et al.* (1972). Further work is required to understand the mechanism of F loss.

The evolution of HCl from hilgardite, $Ca_2B_5O_9Cl{\cdot}H_2O$, (Rachlin *et al.*, 1986), between 950°and 1200°C after the loss of H_2O between 420°and 640°C raises the question of the source of the H that forms the HCl. Some of the HCl possibly formed in a high-temperature reaction between residual H_2 or H_2O and evolved Cl, similar to the reaction of F discussed by Nikolaev *et al.* (1972). Because some F-bearing and Cl-bearing volatiles may be condensable, a sample cannot be considered F- or Cl-free simply because no volatiles were observed by EGA.

Many minerals contain very weakly bonded volatile components, which may be lost by exposure to high vacuum at room temperature. The mineral weddellite, $Ca(C_2O_4){\cdot}2.26H_2O$, is an extreme case. It loses 91 wt. % of its H_2O under such conditions (Mandarino and Witt, 1983). With the ROM instrument, these room-temperature losses could only be determined indirectly as the difference between the TGA loss recorded during heating and the total loss determined by microbalance at the end of the run (i.e., sample wt. - product wt.). If a significant loss takes place at room temperature, a further loss invariably occurs as heating commences, which can be studied by EGA. Presumably these volatiles emitted during initial heating are representative of those lost during pump-down. To date, H_2O has been found to be the predominant loosely held volatile, although other minor components may be present. Small quantities of acetone were identified in low-temperature losses from kraisslite, $(Mn^{2+},Mg)_{24}Zn_3Fe^{3+}(As^{3+}O_3)_2(As^{5+}O_4)_3(SiO_4)_6(OH)_{18}$, but the acetone was most likely an artifact of preparation.

The decomposition of arsenates and arsenites is marked by the detection of $As_2O_3^+$ (and numerous other fragments), as well as $AsCl_2^+$ from Cl-bearing phases. The breakdown of Cl-bearing borates produces BCl_3^+ and $CaCl_4^+$ and their fragments and, possibly, BO^+.

The study of oxalates may be complicated by the formation of elemental carbon during thermoanalysis. If the thermal product of calcium oxalate is a medium gray, due to residual carbon instead of the expected white of CaO, the analysis may be completed by heating the product in O_2 to oxidize the C.

The interpretation of the thermoanalysis of sulfides is assisted by the large volume of literature available on the phase relations of sulfides; however, the phase diagrams represent equilibrium conditions and the thermograms represent results of a dynamic, non-equilibrium system, with the volatiles constantly being pumped from the sample as temperature increases. Thus, the bulk composition may be in a state of continuous change. The interpretation of the TGA-EGA results of millerite, NiS, (Figure 9), using the phase diagrams of Kullerud and Yund (1962) for the system Ni-S in the presence of vapor (Figure 10), is a typical example. Although S is first lost in small amounts at 410°C (not shown), the major losses occur in two steps, with S pressure maxima at 713°C and 1142°C (Figure 9). A comparison of the TGA-EGA data with the phase diagram suggests a path from NiS to $Ni_{3-x}S_2$ + αNiS (field 11 in Figure 10) and $Ni_{3\pm x}S_2$ through the first 713°C weight loss. A liquid component appears just below 800°C (field 4, Figure 10), and the gradual loss from 800° to 1000°C is produced by S being pumped from the vacuum system, which keeps the S vapor pressure of the Ni-S system in disequilibrium. At 880°C the S vapor pressure is at a minimum and the bulk composition is Ni_5S_3. The second major S loss at 1142°C occurs from a melt (field 2 = L + Ni_{ss}, Figure 10). Sulfur was detected as S^+_n, where n = 1 to 6, all apparently being ionic fragments of the S_8 molecule. The persistence of a significant S vapor pressure >1200°C after the second major weight loss suggests that some of the S condensed on the walls of the vacuum system en route to the sensor and was then being pumped off as the S vapor pressure dropped.

Volatiles having very low vapor pressures may condense in the coolest zone en route to the spectrometer and will not be detected. Such condensation has been noted particularly for Na-, Pb-, and B-bearing minerals. Graziani and Lucchesi (1982) described the loss of significant amounts of Na and K from scapolites in the form of NaCl and KCl. To prevent condensation of these chlorides, the vacuum passageway must be heated to several hundred degrees.

conclusively that most of the higher-temperature mass loss was due to O_2.

Similar variations in TGA results for hydrous minerals occur if the experiments are conducted with and without a purge gas or in dry or water-saturated atmospheres. Experiments conducted without a purge gas or in a water-saturated atmosphere will result in significantly higher dehydration temperatures than those obtained with a typical pure N_2 purge. Again, these differences can be attributed to a mass-action effect. Experiments conducted in a vacuum on materials that evolve a gas yield lower reaction temperatures than those obtained at one atmosphere. Finally, the breakdown of Ca-oxalate monohydrate discussed above is an example of TGA in which a purge gas can react with an evolved gas to affect the thermal reactions. Oxidation of CO evolved from dehydrated Ca-oxalate is an exothermic reaction, and this oxidation can significantly increase the sample temperature, thereby apparently decreasing the reaction temperature.

Wendlandt (1986, p. 12) listed several important sample characteristics that can influence the nature and quality of TGA data. In practice, the most important sample parameters are the amount of sample, the particle size of the powdered sample, and the degree of perfection or order/disorder of the sample crystals. Heat of reaction, sample thermal conductivity, and sample packing are also important in some experiments but will not be discussed here (see Keattch and Dollimore, 1975; Wendlandt, 1986). Modern TGA instruments typically accommodate very small sample sizes (<100 mg), and many experiments today employ sample sizes <50 mg. Even though such sample masses are small by analytical standards, the mass of the sample in a sample pan can still affect the TGA results. Generally, reaction temperatures shift to higher values with larger sample mass, particularly in static (no purge) atmospheres. This effect is less significant in dynamic atmospheres. Less common effects related to sample size are those connected with the thermal conductivity of the sample and those connected with the exothermic or endothermic nature of the studied reactions. Large thermal gradients can exist within samples, particularly those having low thermal conductivities. Both this effect and those related to the exothermic or endothermic nature of reactions can be minimized by using as small a sample mass as possible. As discussed above, diffusive transport is potentially important for controlling the reaction rate. Thus, a larger sample size may give rise to slower rates of gas evolution from the powder. In practice, this effect appears to be minor, perhaps due to the fact that the concentration of evolved gases in pores remains relatively constant as sample size increases (see discussion below on analcime dehydration). Although more gas is evolved from a larger sample, it is evolved into a proportionately larger pore volume.

Variations in the particle size of samples can yield small but significant variations in TGA results, as illustrated in Figures 8 and 9 ($\simeq 10°C$ decrease in reaction temperature from largest to smallest particle size for gypsum and chlorite. The differences probably arise from two sources: the change in the diffusion length within

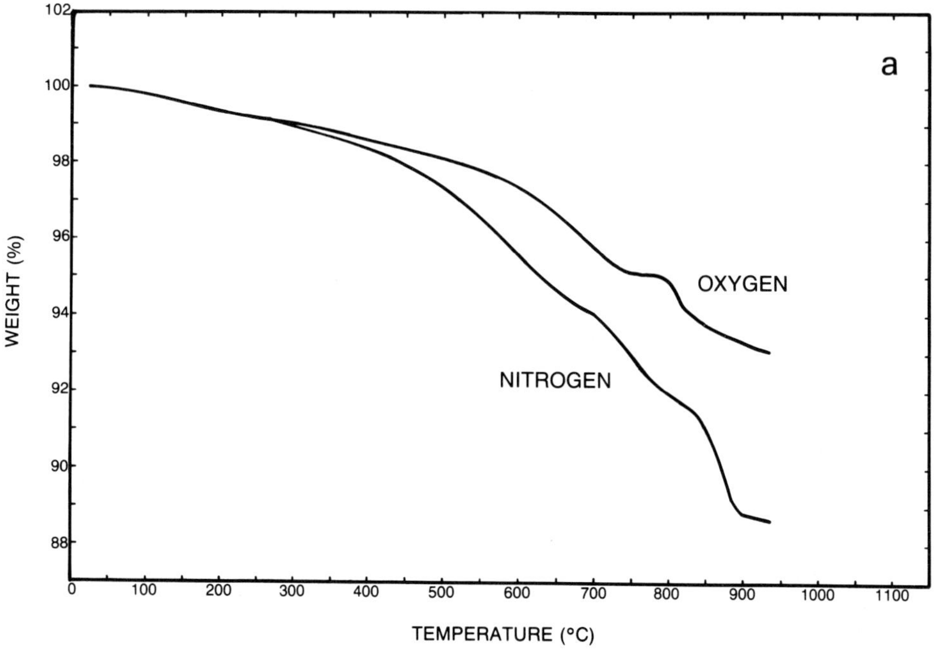

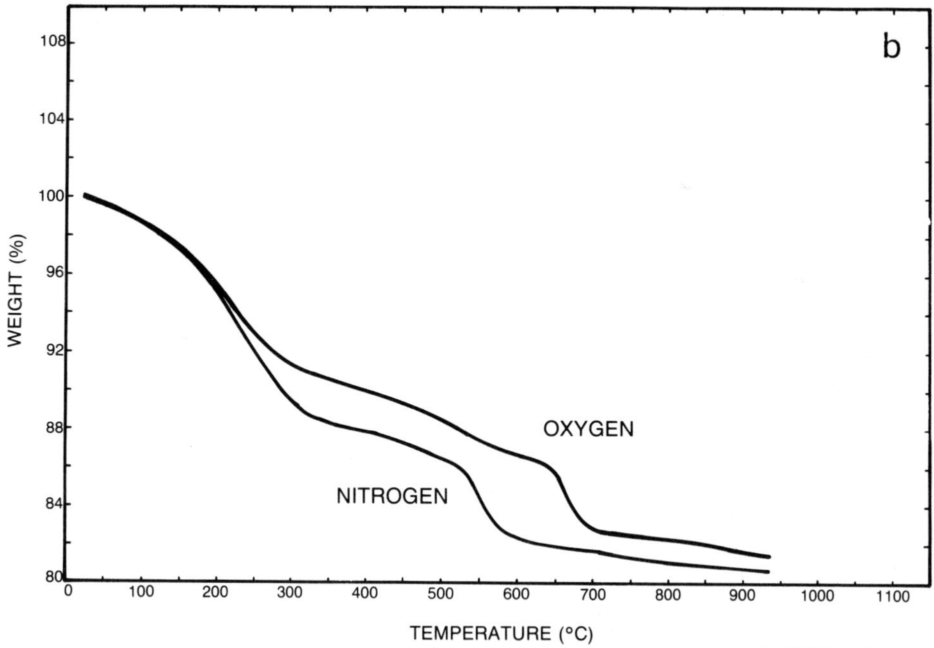

Figure 7. Thermogravimetric analysis curves of (a) romanechite (Van Horne, Texas, HU97618) showing the effects of thermobalance atmosphere on mass-loss curves (O_2 curve, 36.77-mg sample, 10°C/min heating rate, 50-cm^3/min O_2 purge; N_2 curve, 31.60-mg sample, 10°C/min heating rate, 50-cm^3/min N_2 purge); and (b) todorokite (Farragudo, Portugal, HU 106783), showing the effects of thermobalance atmosphere on mass-loss curves (O_2 curve, 12.86-mg sample, 10°C/min heating rate, 50-cm^3/min O_2 purge); N_2 curve, 11.72-mg sample, 10°C/min heating rate, 50-cm^3/min N_2 purge.

MINERAL INDEX

J. W. STUCKI

Alumina 38
Aluminum oxide 34-35, 38, 52, 85
Amphibole 177
Analcime 143-150
Anhydrite 100, 132
Apophyllite 177
Aragonite 119, 122, 132
Arsenates 179, 184
Arsenites 179
Astrophyllite 177

Barium chloride 142
Bentonite 138
Benzoic acid 34, 36, 38
Biotite 177
Boehmite 133-134
Borates 2, 179
Brucite 138, 140, 180
Brushite 168

Calcite 119, 122, 132, 134, 138, 180, 183
Calcium oxalate 100-102, 112, 143, 179
Chlorite 62-63, 90-91, 112, 117, 172
Chlorite, CCa-1 117-118
Clinoptilolite 100, 124-128, 130-131, 148
Cobalt chloride 142
Cyanotrichite 160-161

Dickite 42, 116, 121
Dolomite 119, 123

Feldspar 132

Goethite 136, 141-143
Gypsum 100-101 103-104, 107-112, 114-116, 132

Halloysite 39-40, 42, 44, 116, 119
Hematite 136, 141
Heulandite 125-127, 131
Hilgardite 178
Hydralsite 78
Hydrotalcite 132

Illite 75

Kaolin 46, 63, 151
Kaolinite 34, 42, 50, 54, 75, 79, 85-86, 88, 91, 101-103, 111, 116, 129, 132-135, 138, 140-141
Kaolinite, API #9 120
Kaolinite, intercalated 24-25, 28-29, 35, 41-45
Kaolinite, KGa-1 24, 27, 35, 42-43, 54, 78, 82-83, 120
Kaolinite, KGa-2 103
Kraisslite 179

Lepidomelane 177

Magnesite 119, 123
Magnetite 102
Maricopaite 174-175
Metakaolinite 75, 79, 83, 85-88, 102
Mica 56, 63, 76, 90, 172, 177
Millerite 179, 180-181
Montmorillonite 63-64, 67, 70, 75-76, 86, 88-91, 134, 142
Montmorillonite dehydroxylate 75, 88, 89
Montmorillonite, SAz-1 124, 129
Montmorillonite, STx-1 124, 127
Montmorillonite, SWy-1 54-59, 64-69, 71-73, 76, 78, 80-81, 84, 89-90, 124-128, 130, 132
Muscovite 56, 76, 90

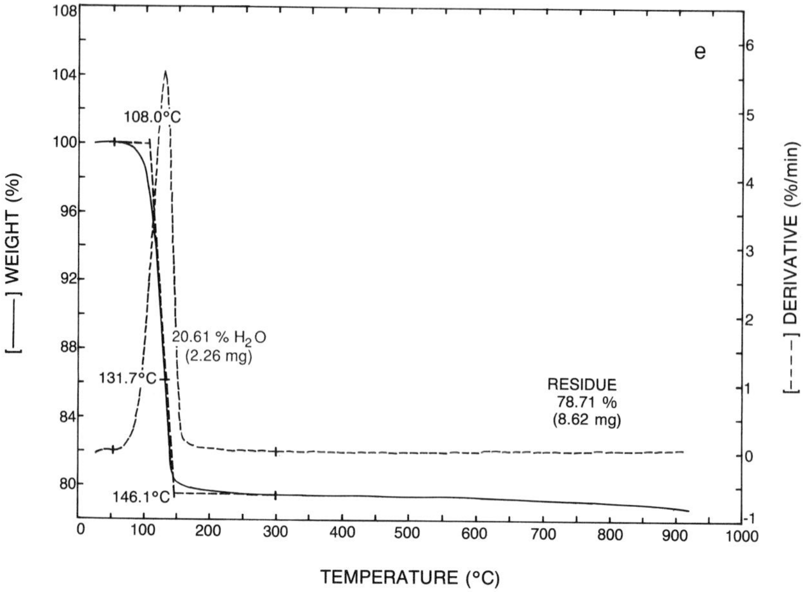

Figure 8e. Thermogravimetric analysis curves of natural gypsum showing the decrease in dehydration temperature with decrease in particle size. (e) <325-mesh sample (<44 μm), 10.95 mg; (10°C/min heating rate, 50-cm^3/min N_2 purge).

the individual particles in a sample and possible structural damage due to grinding. In general, small particle sizes will yield results closer to equilibrium, and reaction temperatures will be lower than for larger particle sizes.

A related factor is the effect of crystal perfection and order/disorder on TGA results. These factors are particularly relevant to studies of clay minerals, which commonly exhibit large variations in crystallinity and order/disorder. The kaolin-group minerals are a good example of these variations, and Figure 10 shows TGA curves for several minerals of this group. The inflection point in the dehydration curve for these samples heated at 10°C/min varied from 484°-495°C for halloysite to 490°-515°C for kaolinite samples to 668°-673°C for dickite samples. These variations reflect the structural differences between halloysite, kaolinite, and dickite and also reflect order/disorder differences among the individual samples. Brindley (1961) discussed the role of crystal structure in dehydration reactions of layer-structure minerals. Smykatz-Kloss (1974) semiquantitatively related the temperature of kaolinite dehydroxylation obtained from DTA to the degree of order of the kaolinite. Clearly, great care must be taken in comparing data obtained in different laboratories on different samples due to the number of instrument- and sample-related factors that can affect TGA results.

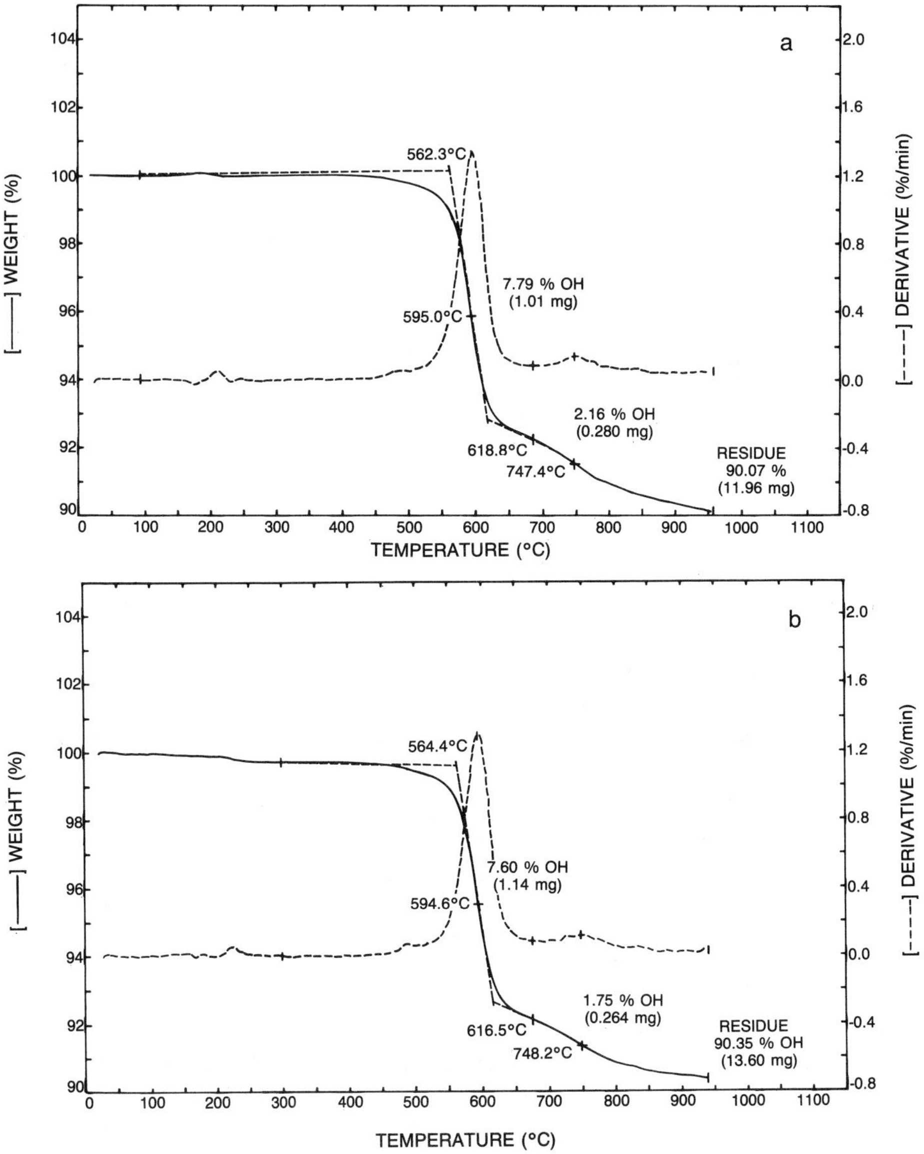

Figure 9a, b. Thermogravimetric analysis curves of ripidolite chlorite, CCa-1, California, showing decrease in dehydroxylation temperature with decrease in particle size: (a) 100-mesh sample (149 μm), 12.99 mg; (b) 200-mesh sample (74 μm), 15.05 mg; (10°C/min heating rate, 50-cm^3/min N_2 purge).

Dollimore, D., Gamlen, G. A., and Taylor, T. J. (1984) Mass spectrometric evolved gas analysis—an overview: *Thermochim. Acta* **75**, 59-69.

Dunner, W. and Eppler, H. (1975) Advanced coupling systems for thermoanalyzer with quadrupole mass spectrometer: in *Thermal Analysis, Proc. 4th Int. Conf. Thermal Analysis, Budapest, 1974, Vol. 3,* I. Buzas, ed., Heyden, London, 1049-1060.

Elliot, R. M. (1969) Single focusing magnetic deflection mass spectrometers: in *Mass Spectrometry, Proc. Symp. Mass Spectrometry, Enfield College of Technology, 1967,* R. Brymner and J. R. Penney, eds., Chemical Publishing Co., New York, 3-14.

Gallagher, P. K. (1982) Applications of evolved gas analysis to the study of inorganic materials and processes: *J. Thermal Anal.* **25**, 7-20.

Graziani, G., and Lucchesi, S. (1982) The thermal behavior of scapolites: *Amer. Mineral.* **67**, 1229-1241.

Green, G. W. (1968) *The Design and Construction of Small Vacuum Systems:* Chapman and Hall, London, 181 pp.

Harris, D. M. (1981) The microdetermination of H_2O, CO_2, and SO_2 in glass using a 1280°C microscope vacuum heating stage, cryopumping, and vapor pressure measurements from 77 to 273 K: *Geochim. Cosmochim. Acta* **45**, 2023-2036.

Hastie, J. W. (1979) Characterization of high temperature vapors and gases: *Proc. 10th Materials Research Symp., Gaithersburg, Maryland, 1978, Vol. 1, Natl. Bur. Stand. Spec. Publ.* **561**, 812 pp.

Hawthorne, F. C., Groat, L. A., Raudsepp, M., Ball, N. A., Kimata, M., Spike, F. D., Lumpkin, G. R., Ewing, R. C., Gregor, R. B., Lytle, F. W., Ercit, T. S., Rossman, G. R., Wicks, F. J., Ramik, R. A., Sherrif, B. L., Fleet, M. E., and McCammon, C. (1990) Alpha-decay damage in natural titanites. *Amer. Mineral.* **75** (in press).

Holdiness, M. R. (1984) Evolved gas analysis by mass spectrometry: A review: *Thermochim. Acta* **75**, 361-399.

Kaisersberger, E. and Emmerich, W. D. (1982) Theoretical and instrumental studies for the coupling of TA systems and quadrupole mass spectrometers: in *Proc. 7th Int. Conf. Thermal Analysis,* B. Miller, ed., Wiley, New York, 279-283.

Karup Moller, S. (1982) Tundrite from the Ilimaussag alkaline intrusion, South Greenland: *N. Jb. Mineral. Mh.*, 481-494.

Kullerud, G. and Yund, R. A. (1962) The Ni-S system and related minerals: *J. Petrology* **3,** 126-175.

Langer, H. G. (1986) Early developments of EGA by mass spectrometry: *Thermochim. Acta* **100,** 187-202.

Langer, H. and Gohlke, R. S. (1967) Mass spectrometric identification of gaseous products from thermal analysis: in *Gas Effluent Analysis,* W. Lodding, ed., Marcel Dekker, New York, 71-100.

Low, M. J. D. (1967) Analysis of gas effluent streams by infrared absorption: in *Gas Effluent Analysis,* W. Lodding, ed., Marcel Dekker, New York, 155-193.

Mackenzie, R. C. and Mitchell, B. D. (1970) Technique: in *Differential Thermal Analysis, Vol. 1,* R. C. Mackenzie, ed., Academic Press, London, 101-122.

Mandarino, J. A. and Witt, N. V. (1983) Weddellite from Biggs, Oregon, U.S.A.: *Can. Mineral.* **21,** 503-508.

Matson, D. W., Muenow, D. W., and Garcia, M. O. (1986) Volatile contents of phlogopite micas from South African kimberlite: *Contrib. Min. Pet.* **93,** 399-408.

McLafferty, F. W., and Stauffer, D. B. (1989) *The Wiley/NBS Registry of Mass Spectral Data,* Wiley, New York, 7 volumes.

Morgan, D. J., Warrington, S. B., and Warne, S. St. J. (1988) Earth science applications of evolved gas analysis: a review: *Thermochim. Acta* **135,** 207-212.

Neumann, B. S. (1977) Thermal techniques: in *Physical Methods in Determinative Mineralogy,* J. Zussman, ed., Academic Press, London, 605-662.

Nikolaev, N. S., Suvorova, S. N., Gurovich, E. I., Peka, I., and Korchemnaya, E. K. (1972) *Analytical Chemistry of Fluorine:* Halsted Press, New York, 222 pp.

Oswald, H. R. and Wiedemann, H. G. (1977) Factors influencing thermoanalytical curves: *J. Thermal Anal.* **12,** 147-168.

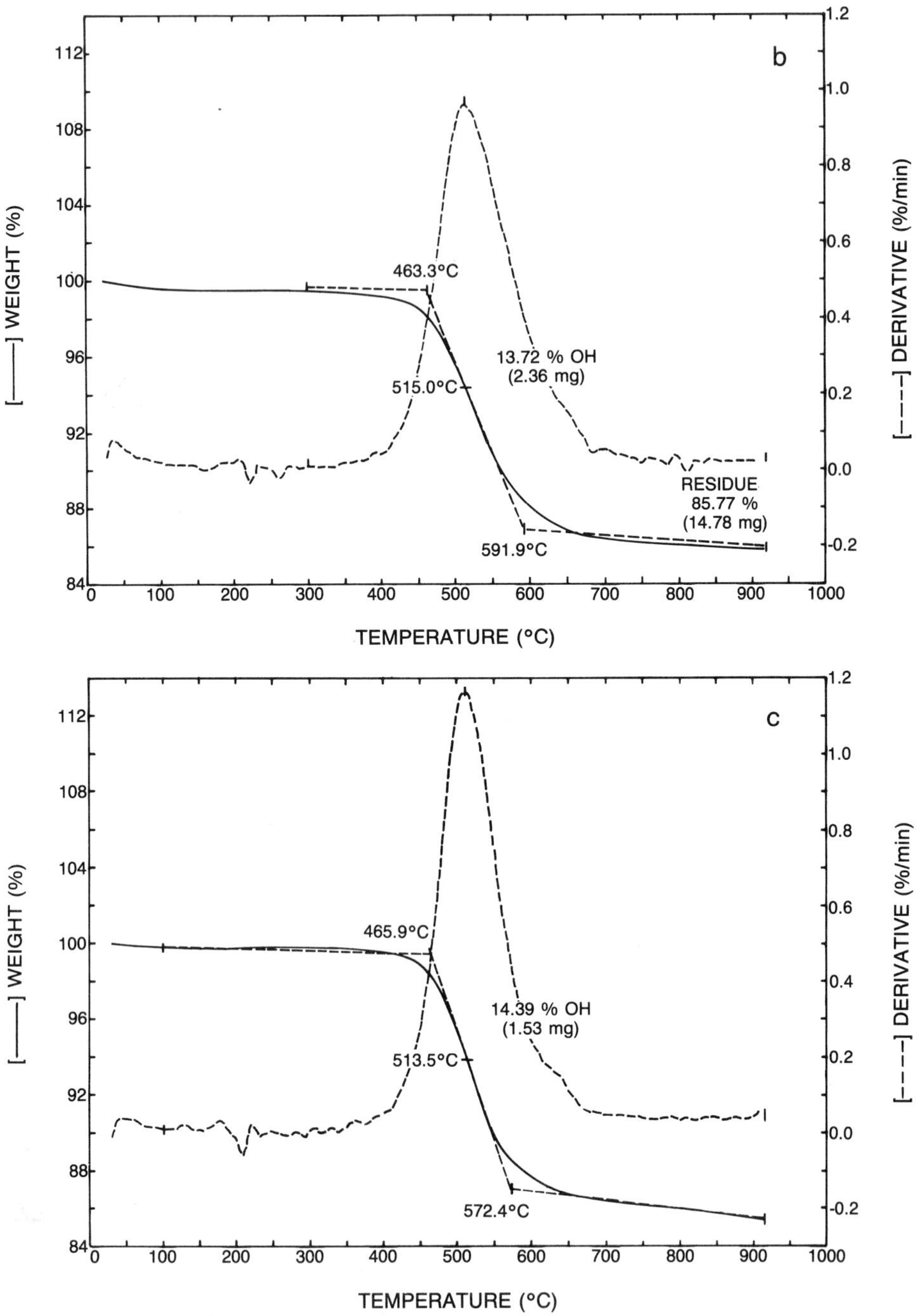

Figure 10b, c. Thermogravimetric analysis curves of kaolin-group minerals showing effects of crystal structure and order/disorder on dehydroxylation: (b) kaolinite, API #9, New Mexico (17.24-mg sample, 325 mesh); (c) kaolinite, KGa-1, Georgia (10.61-mg sample); (10°C/min heating rate, 50-cm^3/min N_2 purge).

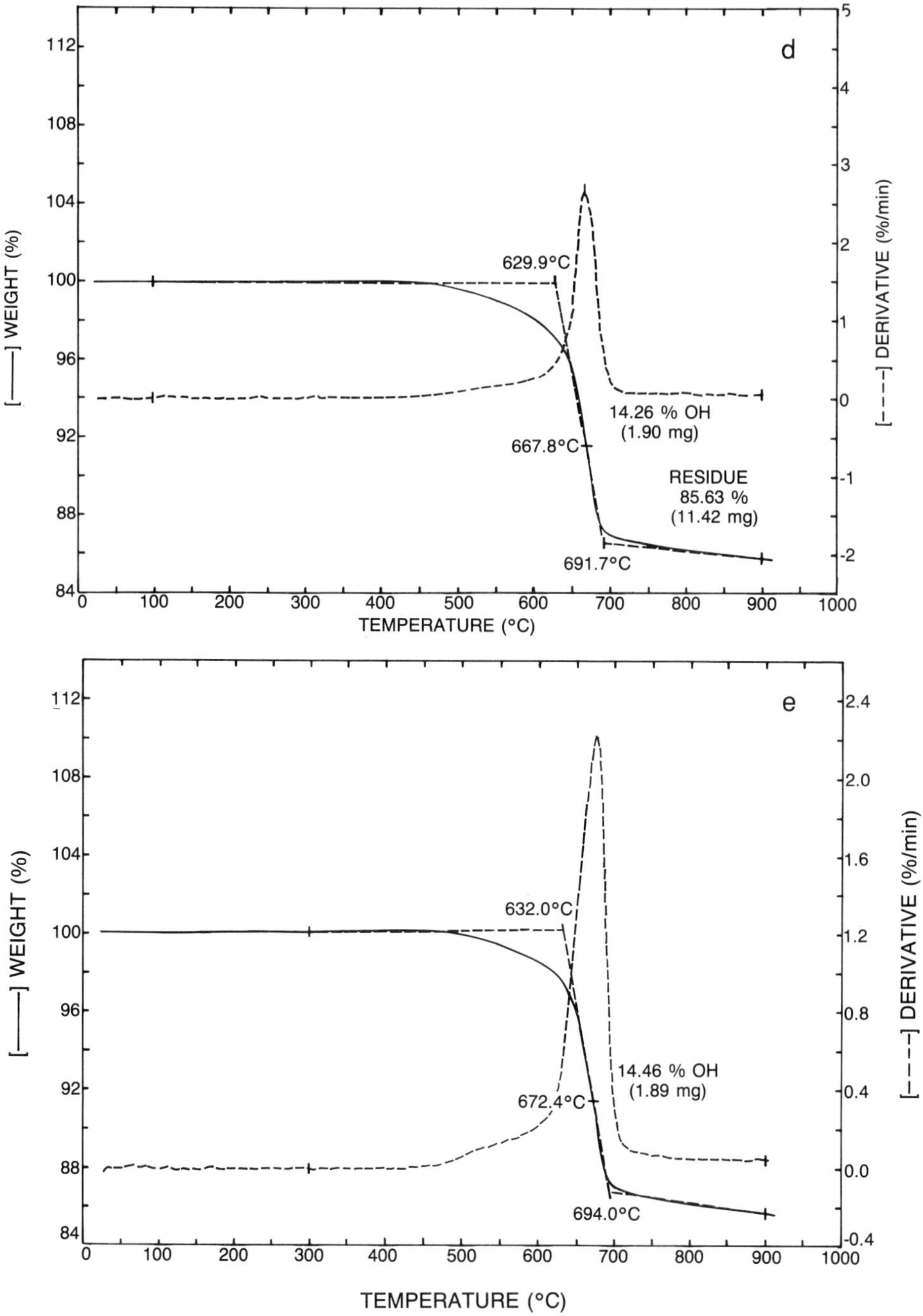

Figure 10d, e. Thermogravimetric analysis curves of kaolin-group minerals showing effects of crystal structure and order/disorder on dehydroxylation: (d) dickite, New Brunswick (13.34-mg sample); (e) dickite, H-14, Colorado (13.06-mg sample); (10°C/min heating rate, 50-cm^3/min N_2 purge).

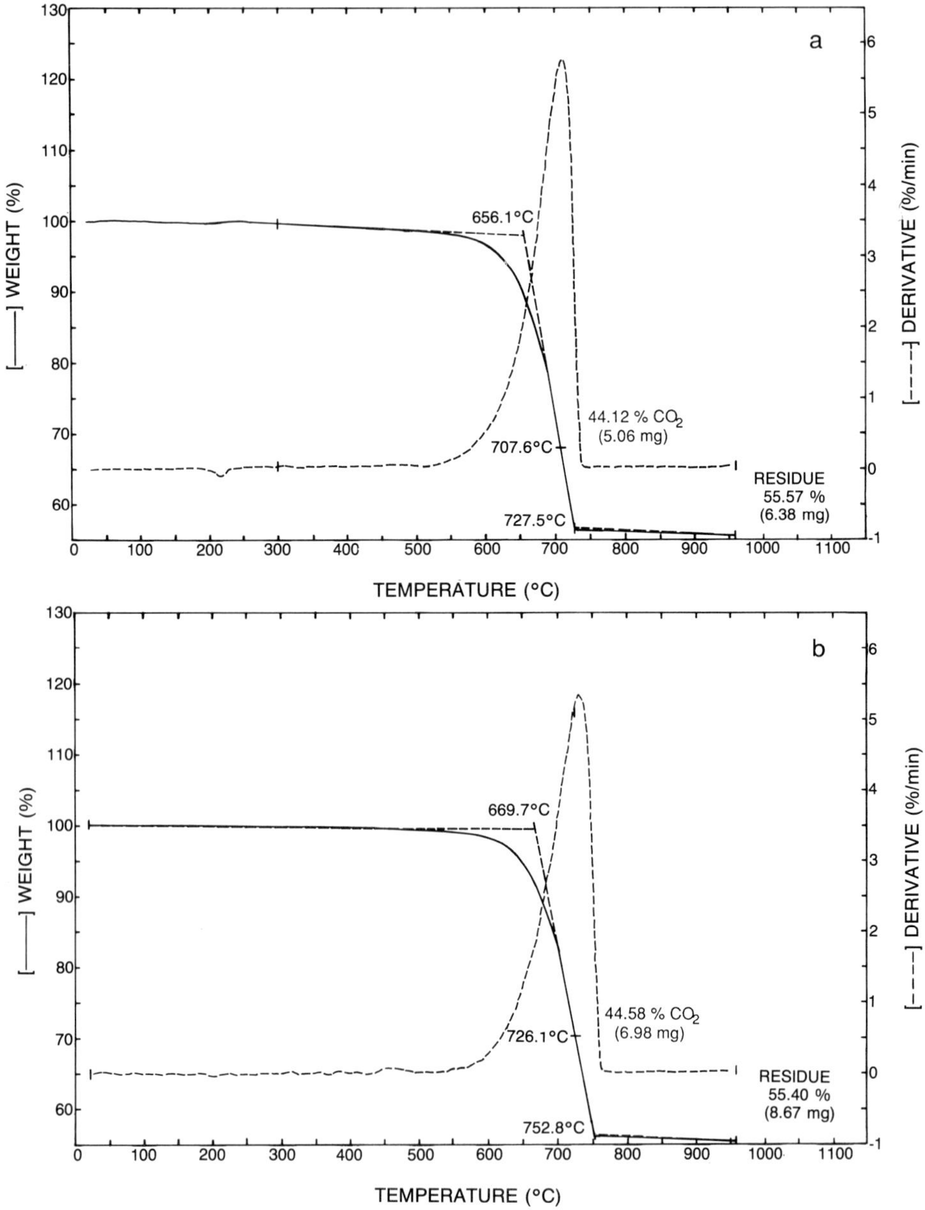

Figure 11a, b. Thermogravimetric analysis curves of carbonate minerals showing effects of crystal structure and composition on decarbonation: (a) calcite, reagent grade (11.48-mg sample); (b) aragonite, synthetic (15.65-mg sample); (10°C/min heating rate, 50-cm^3 /min N_2 purge).

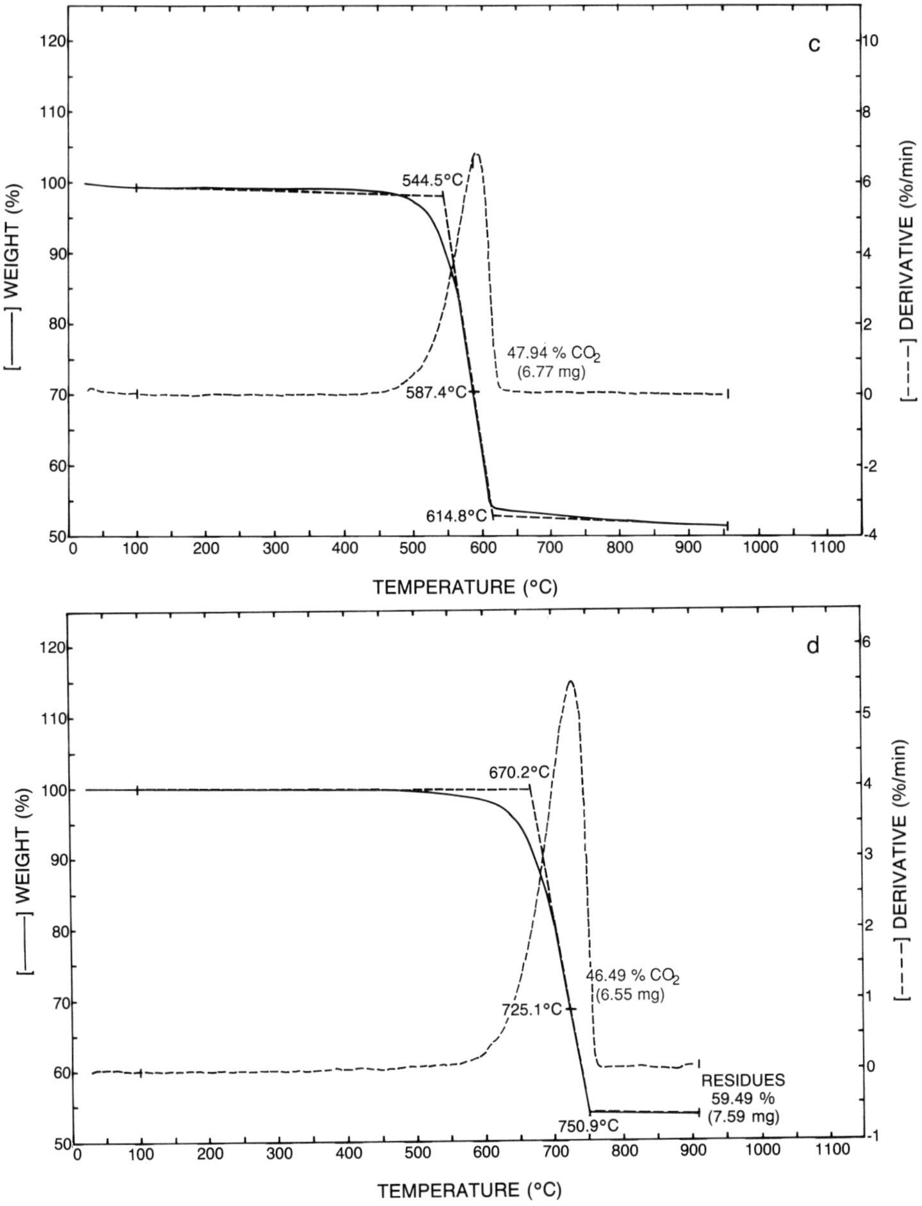

Figure 11c, d. Thermogravimetric analysis curves of carbonate minerals showing effects of crystal structure and composition on decarbonation: (c) magnesite, synthetic (14.13-mg sample); (d) dolomite, Nevada (14.09-mg sample); (10°C/min heating rate, 50-cm^3/min N_2 purge.)

The same energetic principles can probably be applied to dehydroxylation reactions in clay minerals. Giese (1971, 1979) examined hydroxyl orientations in several 2:1 phyllosilicates using an electrostatic model, and these or similar calculations should provide insight into the stability of the OH groups in these minerals. Dioctahedral phyllosilicates typically dehydroxylate at lower temperatures than do trioctahedral phyllosilicates. This stability relation is opposite to that which is normally observed for phyllosilicates. For example, trioctahedral micas weather much more readily than dioctahedral micas (Bassett, 1960), and Giese (1978) showed that dioctahedral minerals usually have greater interlayer bond energies than trioctahedral minerals. The same factors contributing to the lower weathering stability of trioctahedral micas may give rise to higher dehydroxylation temperatures. The more uniform charge distribution around OH groups in trioctahedral minerals may hinder migration of hydrogen and essentially stabilize OH groups.

Thermal analyses reported by various workers (Nutting, 1943; Ross and Hendricks, 1945; Grim, 1953; Grim and Kulbicki, 1961; van Olphen and Fripiat, 1979) demonstrate the thermal behavior of different smectite specimens, but El-Akkad *et al.* (1982) and El-Barawy *et al.* (1986) specifically examined the effects of exchangeable cation on thermal behavior. Data in both papers (Figure 12) show a clear relationship between the exchangeable monovalent cation and the amount of water lost before dehydroxylation (interlayer water), increasing from 6.6% for Cs^+ to 14.5% for Li^+. Data for divalent-cation smectite samples (Figure 13) show a similar trend, although Mg^{2+}-smectite contains less water than would be predicted by compositional trends. These trends are what would be expected based on the sequence of hydration energies for the monovalent and divalent cations and are similar to that observed for zeolites.

Figures 12 and 13 demonstrate that the exchangeable cation has little effect on the dehydroxylation temperature, although a predictable and measurable relation may exist. These TGA data lack sufficient detail to allow multiple water-loss events at temperatures less than that of dehydroxylation to be detected, and DTG curves that might have improved the resolution of these overlapping reactions were not reported. The DTA data presented by El-Akkad *et al.* (1982) and El-Barawy *et al.* (1986), however, demonstrate that the water-loss events at temperatures less than the dehydroxylation temperature are complex for all but the K, Rb, and Cs forms, similar to the results for the zeolite clinoptilolite (see below). TGA curves for smectite reference samples STx-1, SWy-1, and SAz-1 in Figures 14, 15, and 16, respectively, clearly demonstrate a two-stage dehydration reaction and confirm that overlapping reactions are better discerned with DTG than with conventional TGA curves. These data are similar to DTA data reported by Koster van Groos and Guggenheim (1984, 1986, 1987, 1990), who demonstrated that K-, Na-, Mg-, and Ca-exchanged forms of smectite SWy-1 all exhibit two-stage dehydration at ambient and elevated pressures.

Dehydration data similar to the above data on smectites have been reported

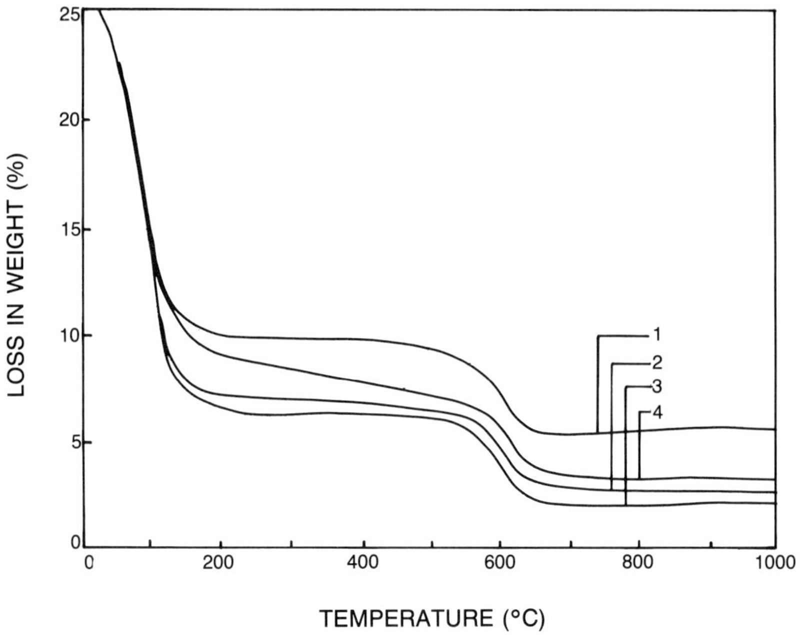

Figure 12. Thermogravimetric analysis curves of monovalent-cation Jelsovy-Potok montmorillonite showing effects of composition on dehydration and total interlayer water content (5 °C/min heating rate, unknown sample masses and purge gas). 1 = Li^+; 2 = Na^+; 3 = K^+; 4 = Rb^+; 5= Cs^+; (from El-Akkad *et al.*, 1982).

for zeolites (Van Reeuwijk, 1974; Simonot-Grange, 1979), and Araya and Dyer (1981) and Bish (1986) reported TGA data for clinoptilolite and heulandite as a function of exchangeable cation composition. Clinoptilolite has an open aluminosilicate structure and possesses several large and distinct three-dimensional cavities. Heulandite is isostructural with clinoptilolite but differs in having a greater Al/Si ratio than clinoptilolite and therefore more exchangeable cations. In addition, heulandite is commonly Ca-rich compared with clinoptilolite. These zeolites exhibit cation-exchange properties similar to those shown by many clay minerals, and many different cation-exchanged forms can be prepared. Considerable insight into the nature of the exchangeable cations and water in the cavities can be gained through thermal analysis.

In contrast to smectites, in which interlayer water sites are all fairly similar, zeolites present a structure in which cations and water molecules can occupy numerous energetically different sites. Traditionally, clinoptilolite has been reported to contain several types of water, including external, loosely bound, and tightly bound or zeolitic water (Knowlton *et al.*, 1981). Figures 17 and 18 show TGA and DTG curves for experimentally exchanged monovalent- and divalent-cation clinoptilolite respectively. Figure 19 shows the TGA curve for two samples of natural, untreated heulandite. As for the smectites, TGA data for the exchanged clinoptilolite samples

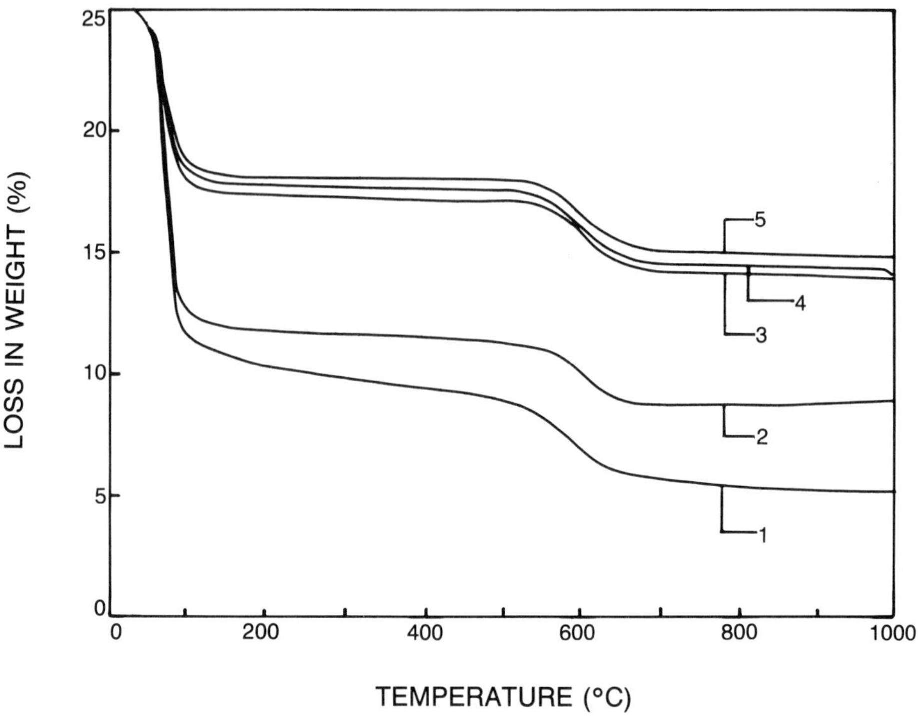

Figure 13. Thermogravimetric analysis curves of divalent-cation Jelsovy-Potok montmorillonite showing effects of composition on dehydration and total interlayer water content (5°C/min heating rate, unknown sample masses and purge gas). 1 = Mg^{2+}; 2 = Ca^{2+}; 3 = Sr^{2+}; 4 = Ba^{2+}; (from El-Akkad *et al.*, 1982).

reveal that the total amount of water lost at temperatures <800°C is strongly dependent on the exchangeable cation. Water losses range from 7.8% for the Cs-clinoptilolite to about 15.5% for Mg-, Li-, and Sr-clinoptilolite samples. Figure 20 shows the dependence of the total water loss on the ionic potential (charge/radius) of the exchangeable cations. All exchanged clinoptilolite samples exhibited major, rapid mass-loss events between 50° and 130°C, with slower and smaller mass losses at higher temperatures. The details of the TGA and DTG curves, however, varied considerably with exchangeable cation. For example, Na-clinoptilolite dehydrated in one broad (perhaps two closely overlapping) step, whereas the Li-, Ca-, and Mg-exchanged samples clearly showed evidence for a complex dehydration process, having as many as three distinct peaks in their DTG curves. In addition, divalent-cation clinoptilolite samples generally retained water to higher temperatures than monovalent-cation samples, a reflection of the greater hydration energy of the divalent exchangeable cations.

Although the zeolite heulandite is isostructural with clinoptilolite, its dehydration behavior is quite different (Figure 19). All heulandite samples examined had

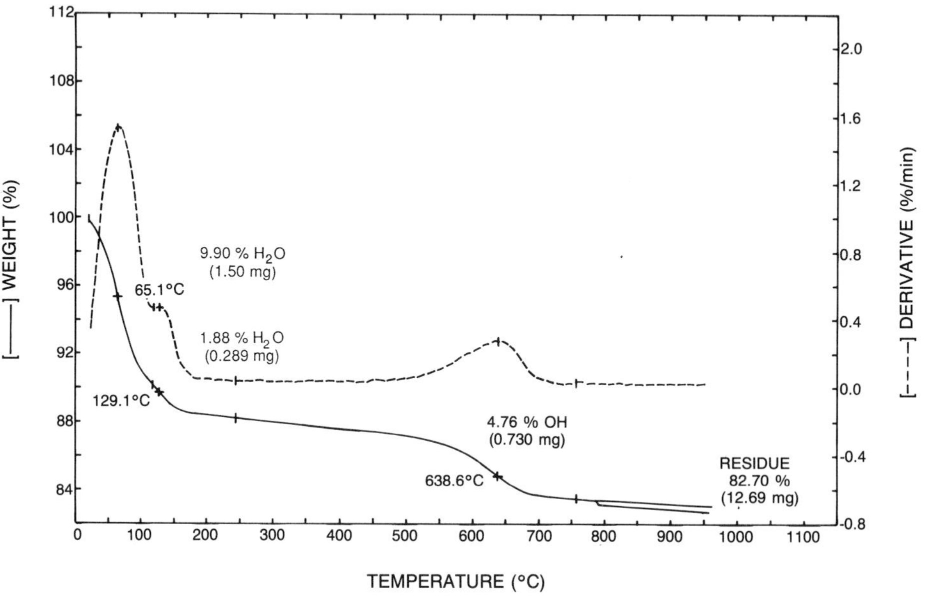

Figure 14. Thermogravimetric analysis curve for smectite STx-1 (Ca-montmorillonite) showing two-stage low-temperature dehydration and dehydroxylation (10°C/min heating rate, 50-cm^3/min N_2 purge, 15.35-mg sample, <0.1 μm particle size).

greater water contents than any clinoptilolite examined and dehydrated in two distinct steps, one up to 180°C (7.7% loss) and one from 180° to 300°C (≈4.0% loss). The latter mass loss, with an extrapolated onset temperature of 205°C, is undoubtedly associated with the formation of heulandite B, reported by Mumpton (1960) to occur at 230°C. The abrupt nature of the second mass loss is likely related to the structural breakdown that accompanies the transition.

The complex dehydration process exhibited by the exchanged clinoptilolite samples was attributed to loss of water associated with individual sites by Araya and Dyer (1981). A more likely scenario, however, is that the multi-stage water loss is due to loss of water associated with different H_2O-cation distances. Water molecules associated with longer H_2O-cation distances, held with less energy, will evolve before the water associated with shorter H_2O-cation distances, held with greater bond energies (Bish, 1986). This relationship is similar to that envisioned for smectite dehydration. During and after the initial water loss, the exchangeable cations move towards the oxygen framework (Galli *et al.*, 1983), and the remaining water molecules are held more tightly. Based on TGA results and structural information tentative conclusions can be drawn about the location of the exchangeable cations and about the details of the dehydration process for the different cation-exchanged clinoptilolite samples. Stepwise dehydration apparently does not result from water completely leaving a cation in one site and then leaving a cation in another site. The de-

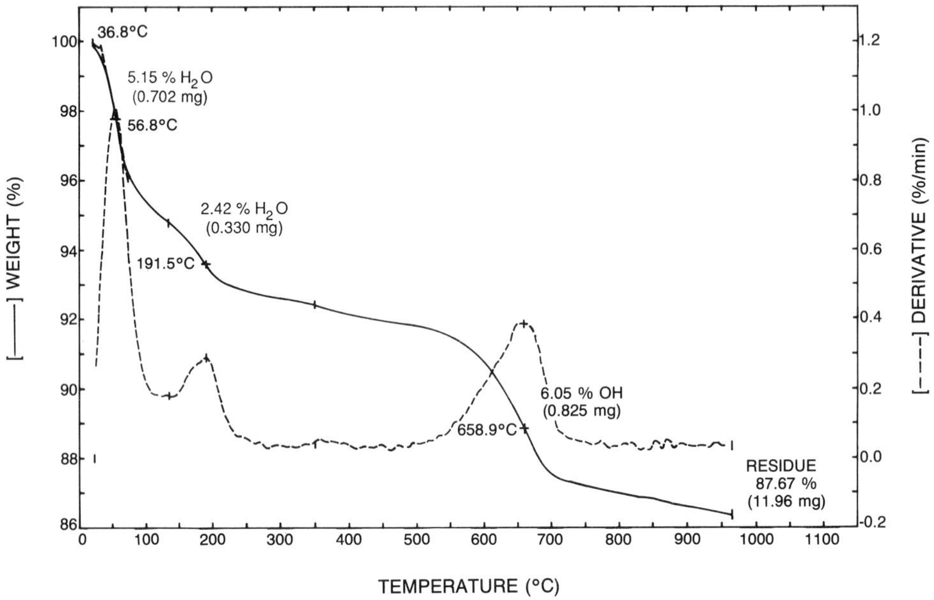

Figure 15. Thermogravimetric analysis curve for smectite SWy-1 (Na-montmorillonite) showing two-stage low-temperature dehydration and dehydroxylation (10°C/min heating rate, 50-cm^3/min N_2 purge, 13.64-mg sample, 0.10-0.35 μm particle size).

hydration behavior of clinoptilolite can be more accurately represented by a continuum of water-molecule binding energies than by a combination of loosely and tightly bound zeolitic water. By coupling TGA and DSC data, the heats of dehydration for these broadly different types of water can be determined (Knowlton *et al.*, 1981).

In addition to delineating the dehydration behavior of clays and zeolites, TGA has been used to elucidate thermal transformations at higher temperatures in phyllosilicates. The work of MacKenzie and Bowden (1983) on amesite is an example of the contribution that TGA can make to the study of thermal reactions. Amesite is significantly different from other 1:1 phyllosilicates in that the development of long basal spacings (≃14 Å) was reported when the mineral had been heated >550°C. Their TGA data, along with DTA and evolved-gas analysis, clearly showed the two-stage nature of the dehydroxylation of this material. X-ray powder diffraction (XRD) results suggested that the first stage of the dehydroxylation reaction was a collapse of some of the material to a pseudo-closed-packed structure, having a long basal spacing. A similar example of the utility of coupled thermal analyses and structural studies of the phyllosilicate minnesotaite was described by MacKenzie *et al.* (1986).

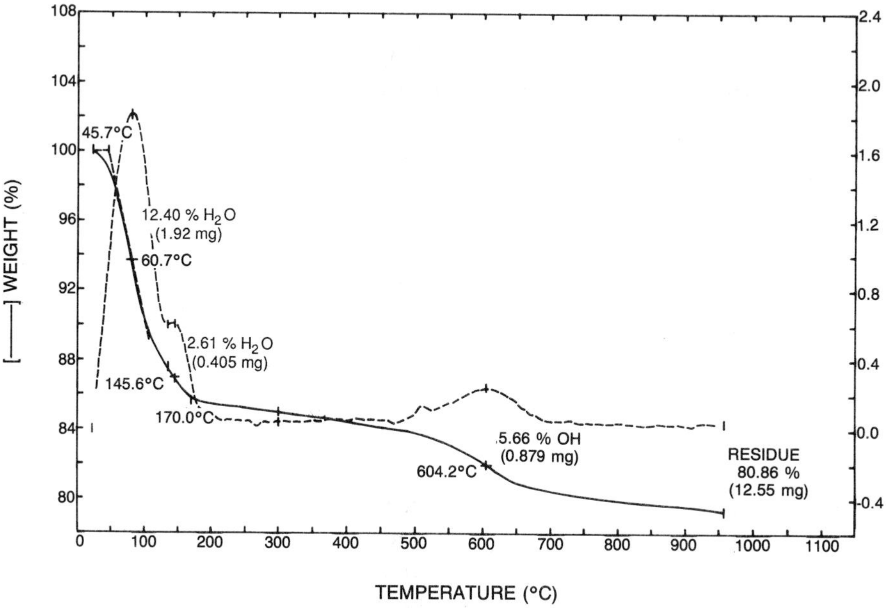

Figure 16. Thermogravimetric analysis curve for smectite SAz-1 (Ca-montmorillonite) showing two-stage low-temperature dehydration and dehydroxylation (10°C/min heating rate, 50-cm^3 /min N_2 purge, 15.52-mg sample, <0.1-μm particle size).

TREATMENT OF THERMOGRAVIMETRIC DATA

In addition to providing information on compositional effects in hydrous minerals, TGA can also give several other useful types of information, including amounts of adsorbed and structural water (to be used in calculating structural formulae), amounts of non-water volatiles (if H_2O is known from other analyses), and quantitative analyses of multicomponent mixtures.

Structural formulae

Silicate clay minerals typically contain both OH^- and molecular H_2O, which are accommodated in relatively specific structural sites. The use of TGA allows the different types of "water" in clay minerals to be differentiated and the relative amounts of structural OH^- and interlayer H_2O to be determined. Careful TGA experiments also allow the relative amounts of first- and second-hydration shell H_2O to be determined in the interlayers of clay minerals (see Figures 14-16). The TGA curve of kaolinite in Figure 4 demonstrates how the OH content of a mineral can be obtained from TGA data. Assuming an ideal anhydrous composition of kaolinite (54.1% SiO_2, 45.90% Al_2O_3), a mass loss from the TGA curve of 14.2% yields a formula of $Al_2Si_2O_5(OH)_4{\cdot}0.04H_2O$. These results differ only slightly from the ideal mass loss of 13.96%, probably due to small amounts of surface-adsorbed water that are lost at lower temperatures.

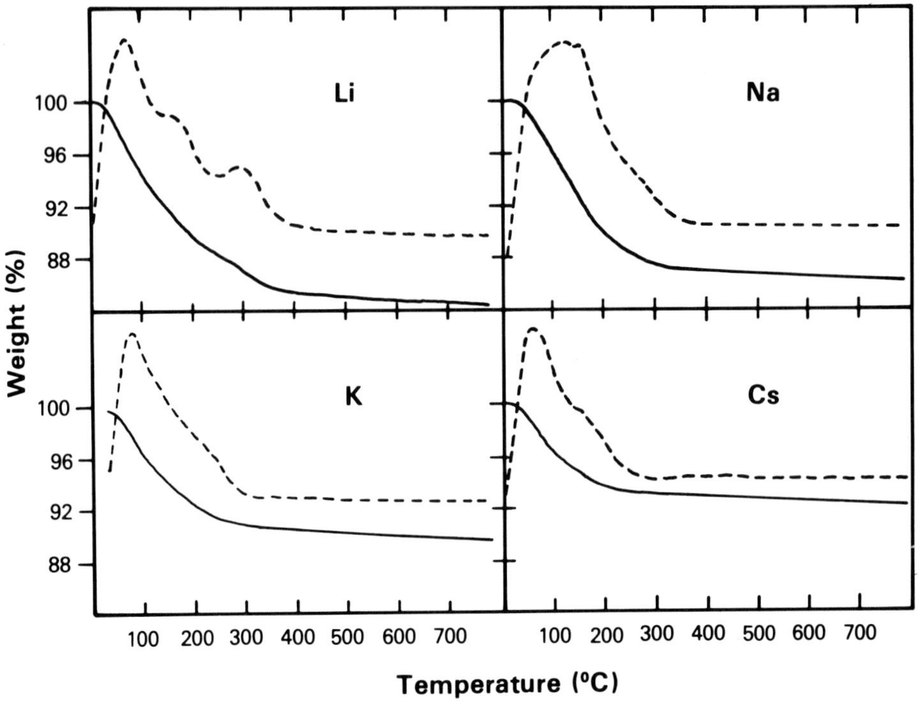

Figure 17. Thermogravimetric analysis curves for monovalent-cation clinoptilolite (#25525, Castle Creek, Idaho). Solid lines are mass-loss curves; dashed lines are derivative curves (10°C/min heating rate, 50-cm^3/min N_2 purge, 10.0-12.5-mg samples.)

Smectites are examples of clay minerals for which both the structural OH^- and interlayer H_2O contents can be determined from TGA. The TGA curves in Figures 14-16 clearly show the loss of interlayer water followed by the loss of structural OH^-. By measuring the mass loss due to these two processes, the amounts of OH^- and H_2O can be determined for use in calculating a structural formula. Determining the break between loss of interlayer H_2O and structural OH^- is not always straightforward, but typically a point on a plateau between the major mass losses is used. This break is particularly difficult to pinpoint for very finely crystalline materials, such as smectites, because molecular water can be held to fairly high temperatures in a defect structure, and structural hydroxyl can begin coming off at relatively low temperatures due to structural perturbations. Using the chemical data in van Olphen and Fripiat (1979) and the TGA data in Figure 15 for smectite SWy-1, the following approximate structural formula is obtained:

$$(Ca_{0.34}Na_{0.19}K_{0.02})(Al_{1.45}Fe^{3+}{}_{0.18}Fe^{2+}{}_{0.01}MgO_{0.27})Si_4O_{10}(OH)_{2.8}\cdot 1.2(H_2O)_1\cdot 0.55(H_2O)_2.$$

This structural formula illustrates some of the difficulties encountered in using TGA data to obtain information on the relative amounts of structural OH^- and interlayer

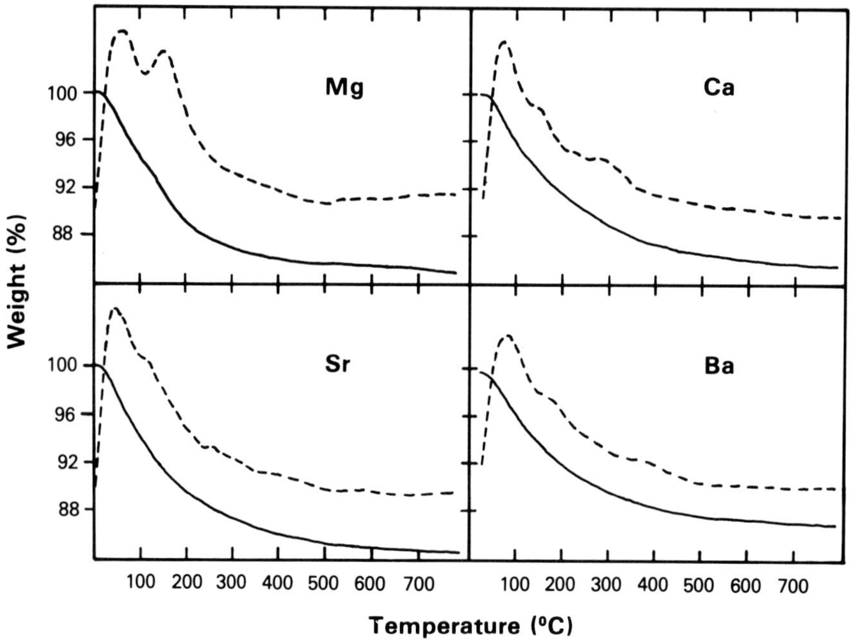

Figure 18. Thermogravimetric analysis curves for divalent-cation clinoptilolite (#25525, Castle Creek, Idaho). Solid lines are mass-loss curves; dashed lines are derivative curves (10°C/min heating rate, 50-cm^3 N_2 purge, 10.0-13.0-mg samples).

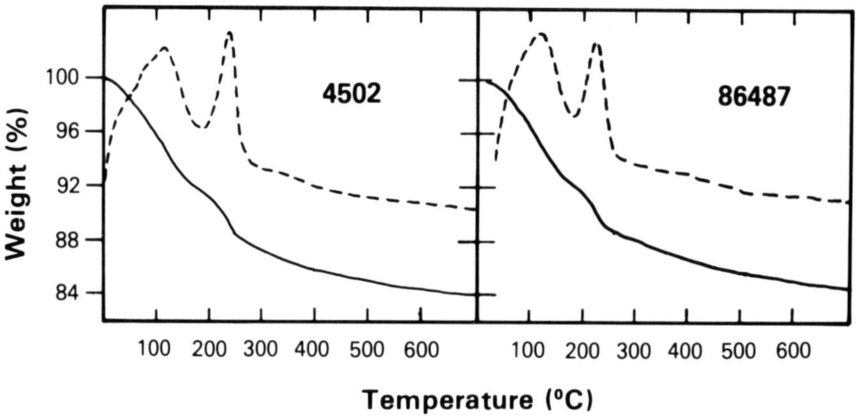

Figure 19. Thermogravimetric analysis curves for heulandite samples (#4502, Nova Scotia; #86487, Faroe Islands). Solid lines are mass-loss curves; dashed lines are derivative curves (10°C/min heating rate, 50-cm^3/min N_2 purge, 11.2-mg samples).

H_2O in clay minerals. The reason for the apparent excess of OH^- (2.8 instead of 2.0) probably lies in the choice of the region on the TGA curve that was thought to represent the dehydroxylation region of the smectite. The region chosen for the hydroxyl loss of sample SWy-1 corresponds roughly to that which would normally be reported as H_2O+, and the mass loss from 350°C to the end of the TGA experiment (6.05%) is similar to the H_2O+ values reported for many montmorillonite samples by Weaver and Pollard (1973, 64-65). Despite the difficulties in assigning mass losses to OH^-, TGA data provide information on interlayer water, which is difficult to obtain by other means.

Analysis for non-water volatiles

Many of the materials clay mineralogists encounter yield mass losses in a TGA experiment that are not due to water evolution (e.g., carbonates and sulfates). Analysis for the volatiles evolved from these minerals is relatively straightforward, and structural formulae can be calculated just as they are for kaolinite. For example, for aragonite and calcite ($CaCO_3$), a mass loss of 43.97% due to the loss of CO_2 would be expected. In addition, knowing the expected mass loss from the breakdown of a specific mineral such as calcite, the amount of that mineral in a mixture can be calculated, as is discussed below. Minerals for which TGA data are particularly useful are those that evolve both water (OH^- and H_2O) and non-water volatiles. Gypsum is an example of such a mineral, although dehydrated gypsum (i.e., anhydrite) does not evolve SO_2 in the temperature range normally employed in TGA. TGA can provide valuable information on mixed hydroxides, such as hydrotalcite, pyroaurite, and takovite. These minerals contain structural OH^-, interlayer H_2O, and interlayer anions, such as CO_3. Coupled with methods specific for H_2O, TGA can provide information on the relative amounts of these volatiles in these phases (Bish and Brindley, 1977; see below). Methods specific to water include the Penfield method and coulometric methods, such as that used by DuPont's Moisture Evolution Analyzer.

Quantitative analysis of mixtures using TGA data

As mentioned above, because many minerals have relatively constant mass losses over well-defined temperature ranges, the relative amounts of these minerals in a multicomponent mixture can be determined using TGA data. As pointed out by Earnest (1980), quantitative analysis ideally requires that the components of a mixture do not have overlapping mass losses. Mixtures of kaolinite with quartz and feldspar, of mixed hydroxides with clay minerals, and of calcite with clay minerals (Figure 21) are good examples of the assemblages for which fairly precise quantitative information on relative amounts can be obtained. Earnest (1980) showed that quantitative analysis of relatively pure kaolinite can be made with TGA data knowing the expected mass loss due to dehydroxylation.

Bish and Brindley (1977) used a combination of XRD data, infrared spectra, and chemical and CO_2 analyses to decipher the chemical composition of the natural Ni-Al mixed hydroxide, takovite. The takovite samples were admixed with different

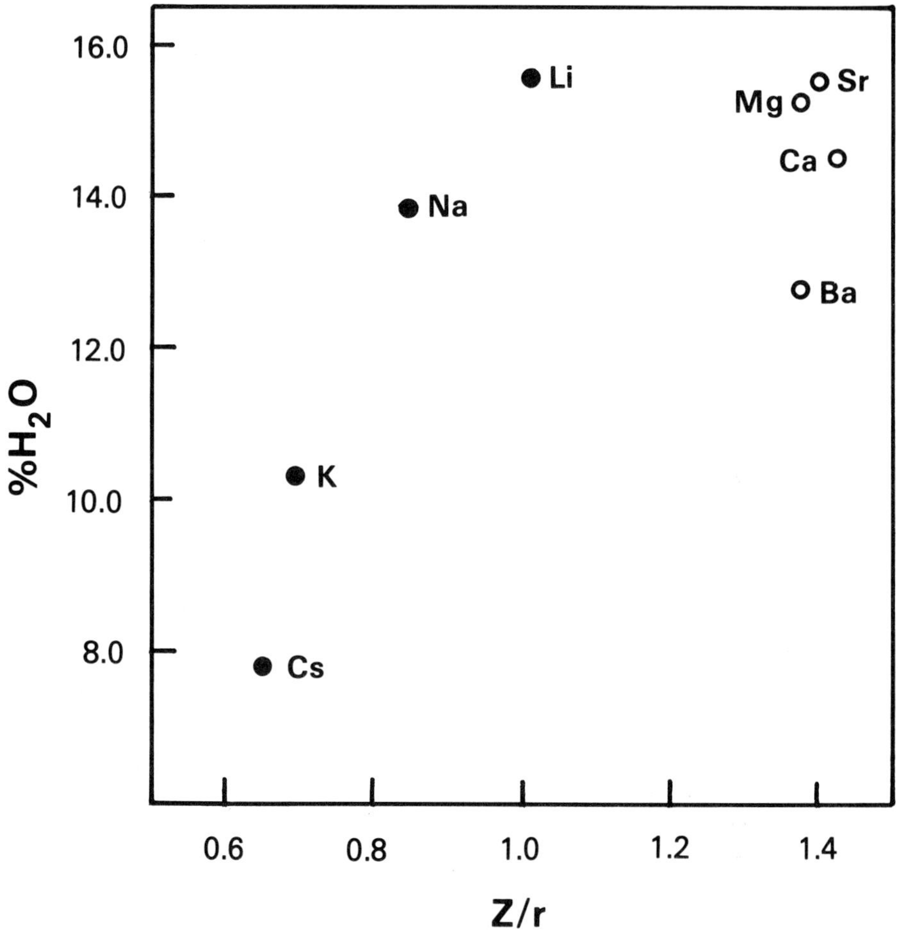

Figure 20. Weight percent water vs. ionic potential, z (charge)/R (VIII-coordinated crystal radii, Shannon, 1976). Ionic potential is weighted according to the observed chemistry.

amounts of kaolinite and boehmite, yielding the TGA curves shown in Figure 22. For samples (a) and (b), W_1 represents the loss of molecular water from takovite. The second major step in the TGA curves corresponds to the loss of CO_2 and OH^-, which is largely complete at 400°C. The less well defined losses in the 350° to 600°C range are due to loss of water from the kaolinite and boehmite impurities. For sample (a), in which kaolinite appears to be the only significant impurity, the SiO_2 content (14.04%) from chemical analysis corresponds to 30.2% kaolinite and 4.21% H_2O. A mass loss of this magnitude is marked K in the figure. Thus, the total mass loss for sample (a), W_2 = 18.7%, is the loss due to CO_2 and OH^- from takovite plus 4.2% from the kaolinite impurity. Therefore, the amount of CO_2 plus OH^- in tako-

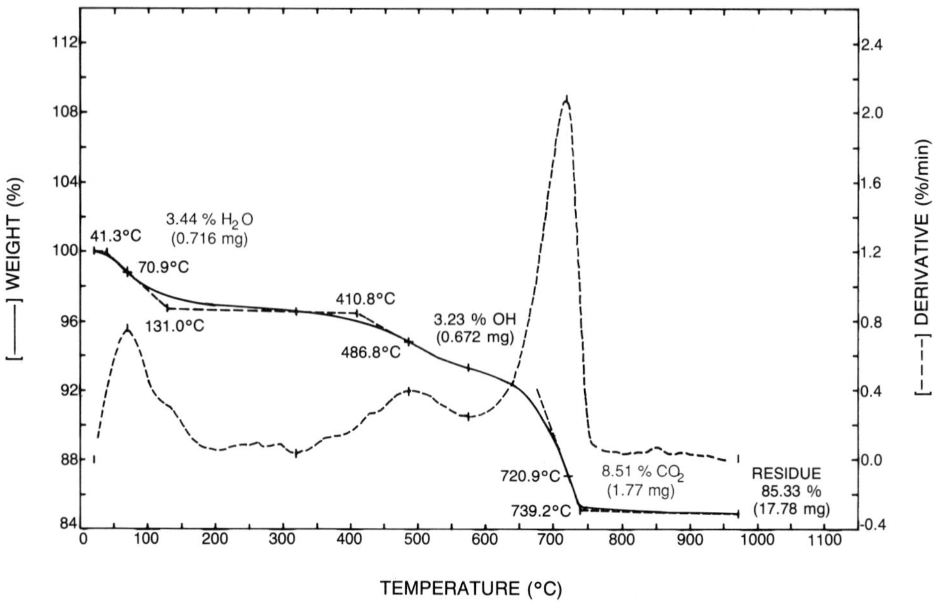

Figure 21. Thermogravimetric analysis curve of brick component from Acme Brick Company, showing mass losses due to dehydration of smectite, dehydroxylation of kaolinite and smectite, and decarbonation of calcite (20°C/min heating rate, 50-cm^3/min N_2 purge, 20.83-mg sample).

vite is 14.5%. For sample (b), the mass loss marked B + K in Figure 22 arises from boehmite and kaolinite. The SiO_2 content from chemical analysis (3.72%) corresponds to 8.0% kaolinite and 1.12% H_2O. The total mass loss marked B + K is 5.8%, of which 1.1% is attributed to kaolinite and the remainder attributed to boehmite, corresponding to 31.5% boehmite. Therefore, W_2 for sample (b), 19.1%, is due to the loss of CO_2 and OH^- from takovite plus 5.8% water from kaolinite and boehmite. Thus, the amount of CO_2 plus OH^- in takovite is 13.3%. The derived amount of boehmite plus kaolinite can subsequently be used to correct the chemical analyses to determine the structural formulae of the takovite samples, a task that would be difficult by other means.

In another application, Figure 21 shows the TGA curve for a raw material used in brick making. From semiquantitative analysis using XRD data, this material contains about 28% quartz, 25% calcite, 7% kaolinite, and 40% smectite. The mass losses in Figure 21 at temperatures <200°C are due to smectite dehydration, that at about 500°C is due to kaolinite dehydroxylation, and the large loss at temperatures >600°C is due to a combination of calcite breakdown and smectite dehydroxylation. Thus, XRD and TGA data can be combined to give a better estimate of the amounts of phases in this sample, knowing the expected mass losses from kaolinite, smectite, and calcite.

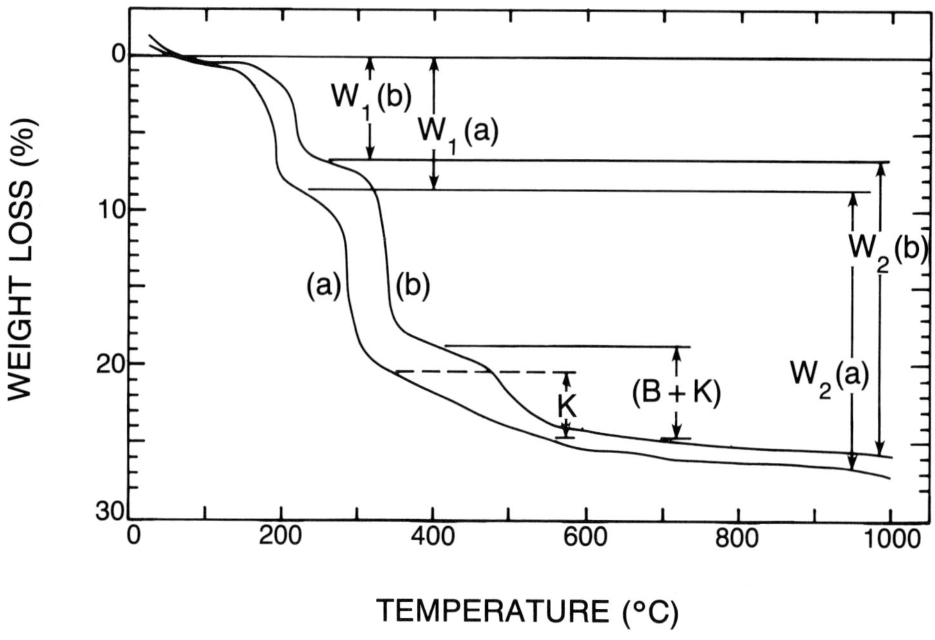

Figure 22. Thermogravimetric analysis curves of takovite from (a) Australia and (b) Var, France. W_1 = mass loss due to molecular water; W_2 = combined higher temperature mass loss; K = mass loss due to kaolinite impurity; B + K = mass loss due to boehmite + kaolinite impurities (5 °C/min heating rate, no purge, 10-mg samples; from Bish and Brindley, 1977).

Johnson *et al.* (1985) described a method using simultaneous linear equations for quantitative analysis of mixtures that employs mass-loss data in addition to other types of quantitative data, including XRD and chemical analyses. Theoretically, this approach allows TGA data for mixtures to be employed, even if the component phases have overlapping mass losses. For phases having well-defined mass losses, this approach using TGA data coupled with other data (e.g., XRD data) should yield very accurate results. The above TGA data for the brick raw material are a good example of a system that should be amenable to solution by this coupled method.

ADVANCED TREATMENT OF TGA DATA

In addition to using TGA as an analytical tool to provide information on the composition of materials and mixtures, TGA data can be used to study the details of many thermal reactions involving mass losses, including the kinetics of the reactions and determination of equilibrium dehydration behavior.

Determination of reaction kinetics from TGA data

Thermogravimetric analysis can be used to evaluate the kinetics of any solid-state reaction that involves either mass loss or gain, and a very large body of literature exists concerning reaction kinetics obtained from TGA data. Two distinct ways exist for determining the kinetics of thermal reactions: isothermal methods and dynamic methods. The study of reaction kinetics by conventional isothermal methods involves either classical treatment of heating-experiment data or, in a few instances, evaluation of the data with time-temperature-transformation (TTT) theory (Putnis and McConnell, 1980). Nonisothermal methods of obtaining kinetic information rely on the fact that one dynamic TGA curve is equivalent to several isothermal experiments. The literature on nonisothermal kinetic evaluation methods is voluminous, and numerous methods have been described to study kinetics. For this reason, some of the more common methods of nonisothermal analysis are summarized here; references are provided for the interested reader. Although considerably more time-consuming, isothermal methods of analysis suffer from fewer problems of data analysis and have distinct benefits in evaluating reaction order. As emphasized by Goss (1987), however, any kinetic treatment cannot completely define solid-state transformations because of the empirical nature of the kinetic theory. Therefore, any complete kinetic study, isothermal or nonisothermal, should include information on mechanisms of transformation, provided, perhaps, by microstructural studies.

Nonisothermal reaction kinetics. Nonisothermal methods of kinetics analysis have been used for more than 35 years to provide information on a variety of systems, notably those involving polymers and the pyrolysis of organic materials. Numerous advantages and disadvantages exist for this method of obtaining reaction kinetics. The advantages include: (1) the fact that a single dynamic TGA experiment can provide considerably more data than an isothermal run, hence, far fewer experiments are required; (2) kinetics can be examined over a large temperature range; (3) a single sample is sufficient to study the entire temperature range, but multiple samples are required for isothermal studies; (4) the reaction onset temperature can be measured precisely; (5) sample reaction during heatup is not a problem in nonisothermal studies, whereas a sample can undergo considerable reaction during heating to the chosen reaction temperature in isothermal studies (Coats and Redfern, 1963; Bagchi and Sen, 1983).

Several authors, notably Arnold *et al.* (1979), have discussed the problems inherent in evaluating reaction kinetics using nonisothermal data, including the experimental conditions that affect TGA data, such as sample size, sample packing, heating rate, amount and composition of the gas present, and the occurrence of endothermic or exothermic reactions within the sample. In a thorough study of the goethite-to-hematite reaction, Goss (1987) discussed a variety of problems inherent in the analysis of reaction kinetics. Solid-state reactions are typically complex and may proceed through a series of steps. The mechanisms controlling the overall reaction, such as nucleation, growth, or diffusion, will each have a particular activation energy. Thus, the empirical activation energy obtained for the overall reaction will result from a combination of all of the basic mechanisms contributing to the reaction or from one or more dominant mechanisms. Commonly, one reaction in a series of reactions is significantly slower than all other reactions and will thus dominate the kinetics of the overall reaction (rate-controlling step). Dynamic TGA data have a tendency to contribute to an overlapping of reactions, leading to, for example, an overlapping of derivative curves. Both isothermal and nonisothermal kinetics analyses demand that particular attention be paid to the existence of independent, perhaps overlapping, reactions, and that they be treated separately. Finally, although overlapping reactions can be separated using nonisothermal treatments, these nonisothermal methods are often inappropriate or difficult to apply to the determination of reaction mechanism or order of reaction.

Briefly, the basic equation underlying the application of dynamic TGA data to kinetic studies is the same as that used above to describe the processes occurring within a sample in a TGA experiment, namely,

$$-dX/dt = kX^n, \tag{11}$$

where t is time, X is the amount of sample reacting, n is the order of the reaction, and k is the rate constant for the reaction. The rate of reaction is often expressed similarly as,

$$d\alpha/dt = kf(\alpha), \tag{12}$$

where α is the reacted fraction at time t. The term $f(\alpha)$ denotes that the rate of reaction, $d\alpha/dt$, is some function of α; $f(\alpha)$ is commonly used for the derivative function. The temperature dependence of the rate constant, k, is formulated in the familiar Arrhenius equation as follows:

$$k = Ae^{-E/RT}, \tag{13}$$

where A is the pre-exponential factor, E is the activation energy, and R is the gas constant. Kinetic methods employing nonisothermal TGA data may be classified as either derivative, integral, or approximate, depending on the way in which Eq. (8) is treated. Using the following equation from Sestak (1966),

$$-dX = (m_0/w_\infty)dw, \tag{14}$$

where m_0 is the initial mass of the sample and w_∞ is the maximum mass loss, derivative methods can be obtained. Integrating the left side of Eq. (14) from m_0 to X and integrating the right side from 0.0 to w, produces the following equation:

$$X = (m_0/w_\infty)(w_\infty - w). \tag{15}$$

Substituting Eqs. (15) and (13) into the logarithmic form of Eq. (8) and differentiating yields one of the differential methods (see Carroll and Manche, 1972). Integral methods use the integrated form of Eq. (8), and the resultant exponential integrals can be solved by using (1) simple approximation, (2) a series expansion, or (3) tabulated values (Carroll and Manche, 1972).

One of the primary drawbacks to nonisothermal kinetics analysis is that the mechanism of reaction is difficult to deduce. The functional dependence of the reacted fraction and time on the rate-controlling process was discussed by Satava (1971). The process discussed included one-, two-, and three-dimensional diffusion, random nucleation, and phase-boundary reactions. Typically, the sensitivity of these methods for mechanism determination is only fair, but Bagchi and Sen (1983) argued that a combined integral and differential method is highly sensitive to the reaction mechanism. As emphasized above (Goss, 1987), kinetic data should be supplemented with other data whenever possible to allow accurate deduction of the reaction mechanism.

Numerous methods of obtaining kinetic information using the above treatments have been published, many of which are referenced in Sestak (1966), Carroll and Manche (1972), and Wendlandt (1986). Although numerous nonisothermal kinetic analyses of simple inorganic compounds exist, few such studies are of minerals. Wendlandt (1986) discussed nonisothermal kinetic analyses of several phases that occur as minerals, including brucite ($Mg(OH)_2$) and calcite ($CaCO_3$).

Cabrera and Eddleston (1983) examined the nonisothermal kinetics of the dehydroxylation of kaolinite using several approaches and concluded that the dehydroxylation reaction is second order. In addition, they showed that the activation energy for dehydroxylation is a function of kaolinite crystallinity, the more highly crystalline samples having the greater activation energy. The results of Dondur and Vucelic (1983a, 1983b, 1983c) on synthetic zeolites illustrate the application of TGA kinetics to the study of complex dehydration. On the basis of data collected by both nonisothermal and isothermal methods, they discussed the difficulties encountered due to the existence of multiple zeolite-water complexes. Their approach to solving this complex dehydration process was to treat the system as a sum of elementary reactions having an exponential distribution of activation energies. This approach should be equally applicable to clay and zeolite systems having similar complex dehydration behavior.

Isothermal reaction kinetics. In contrast to the use of nonisothermal TGA kinetics analysis, isothermal TGA data have been used in several studies of mineral reactions. These studies typically employ classical kinetic theory; but TTT theory (Putnis and McConnell, 1980) has been successfully employed by some investigators (e.g., Onike *et al.*, 1986) and shows promise in the evaluation of TGA, DTA, and DSC results. One of the first studies of the kinetics of thermal reactions of clays was by Murray and White (1949), who made a cursory examination of several kaolin-group minerals and a Wyoming bentonite. They emphasized the importance of heating rate and water vapor pressure and made the important assumption (probably correct) that P_{H_2O} within their samples during dehydration was about one atmosphere. Most kinetic treatments of isothermal TGA data use the general rate equation (Eq. (12)) and its integrated form, viz.,

$$g(\alpha) = kt, \tag{16}$$

where $g(\alpha)$ denotes some integral function of "α" just as $f(\alpha)$ above denoted some derivative function of "α". Studies to determine mechanisms of reaction rely on a dependence of $f(\alpha)$ or $g(\alpha)$ on reaction mechanism. The forms of $g(\alpha)$ corresponding to a variety of probable reaction mechanisms were derived by Satava (1971), and the mechanism is usually inferred from the form of $g(\alpha)$ that gives the best fit to the experimental data. Complex reactions involving domain formation and phase-boundary advancement, however, require empirical solutions to the rate equation (Redfern, 1987).

Many methods have been suggested for obtaining kinetic information from isothermal experiments (e.g., Keattch and Dollimore, 1975), but only some of the approaches used in the studies of minerals will be described here. Following Hancock and Sharp (1972) and Redfern (1987), the isothermal kinetics of a variety of solid-state reactions can be described by the empirical equation,

$$d\alpha/dt = k^m t^{m-1}(1 - \alpha), \tag{17}$$

where m is a constant that depends on the geometry of the system. If k and m are independent of α, the integrated rate equation is

$$-\ln(1 - \alpha) = kt^m \tag{18}$$

or

$$\alpha = 1 - \exp(-kt^m). \tag{19}$$

Eq. (19) is known as the Avrami or Johnson-Mehl equation (Hancock and Sharp, 1972) and is often expressed in the logarithmic form,

$$-\ln[\ln(1 - \alpha)] = \ln k + m \ln t. \tag{20}$$

If the reaction being studied obeys Eq. (19), plots of $\ln\ln(1 - \alpha)$ vs. ln t (so-called "ln-ln plots") will yield approximately straight lines in the range of α from 0.15 to 0.5. Comparison of the slopes of these lines with the values of m for different reactions in Hancock and Sharp (1972) can give some indication as to the reaction mechanism. At higher temperatures, however, many thermal reactions do not yield slopes corresponding to any of these mechanisms, and the reactions may be heat- and mass-transfer controlled due to high reaction rates (Redfern, 1987). Data for different temperatures that yield ln-ln plots having the same or similar values of m are termed isokinetic, i.e., the reaction mechanism is the same over the temperature range of the data. Hancock and Sharp (1972) suggested several alternative treatments for data not obeying Eq. (19), and numerous other methods of analysis were discussed by Keattch and Dollimore (1975).

Most reactions have temperature-dependent rate constants obeying Eq. (13), with activation energies and pre-exponential factors associated with an elementary event. If E and A are associated with a complex reaction, they are purely empirical and correspond to multiple events, such as nucleation, diffusion, and growth (Goss, 1987; Redfern, 1987). Numerous methods have been employed to determine E and A, and one of the most common methods uses the logarithmic form of the Arrhenius equation,

$$\ln k = \ln A - E/RT. \tag{21}$$

Using values of k determined at several temperatures from the appropriate integrated rate equations, $g(\alpha)$, a plot of ln k vs. 1/T yields values of E and A. If E and A are constant, the slope of the curves is -E/R, and the intercept is ln A. This approach will not reveal changes in E with α, and E may also depend on the integrated rate equation used to obtain k. Hence, more complex methods independent of the empirical rate equations must be employed to indicate any variation of E during reaction. Goss (1987) and Redfern (1987) discussed a method of independently determining E, which involves integrating a non-specific reaction function $g(\alpha)$ from $\alpha = 0$ to Y. The value of the integral is constant with temperature, viz.,

$$\ln t_Y = C - \ln A + E/RT, \tag{22}$$

where C is a constant and t_Y is the time to reach $\alpha = Y$. In this "time-to-a-given-fraction" method, a plot of ln t_Y vs. 1/T has a slope of E/R, if E and A are independent of temperature. Goss (1987) described the similar "change-in-rate" method, which is related directly to the Arrhenius equation,

$$\ln(dy/dt_Y) = \text{constant} + \ln A - E/RT. \tag{23}$$

A plot of the reaction rate, dy/dt, at a particular value of α vs. 1/T will yield activation energies. This method is similar to the "time-to-a-given-fraction" method but overcomes the difficulty in determining the zero point of a reaction, i.e., t = 0 and y = 0.

Hancock and Sharp (1972) compared the kinetics of several isothermal solid-state reactions based on an analysis of nucleation and growth processes in order to distinguish theoretical mechanisms. They applied their method to the decomposition of kaolinite and brucite with some success. Using the "ln-ln" plots discussed above, they showed that limited TGA results for kaolinite yielded a slope, m, of 0.56, suggesting a diffusion-controlled reaction in agreement with results obtained by Brindley *et al.* (1967a). Similar analysis of brucite dehydroxylation yielded a slope of 1.05, suggesting a phase-boundary controlled reaction.

Redfern (1987) employed similar methods in a detailed study of the kinetics of dehydroxylation of kaolinite. A ln-ln plot of his data collected at 461°, 483°, and 503°C revealed that the reactions at these temperatures were isokinetic and suggested a reaction mechanism intermediate between diffusion controlled and first order. Data obtained at 530°C corresponded to a first-order model, and data from 617°C suggested that the reaction was heat- and mass-transfer controlled. Plots of ln k vs. 1/T for data collected at 461°, 503°, and 530°C were similar for both mechanisms, and both yielded activation energies of 222 kJ/mole. Use of the "time-to-a-given-fraction" method revealed that results at 461° to 503°C approached isokinetic behavior. Using groups of data to calculate the evolution of activation energies suggested that the average activation energy increased until $\alpha = 0.4$, at which point it remained relatively constant.

Redfern (1987) inferred from these results that the dehydroxylation reaction starts with a low activation energy and quickly (before $\alpha = 0.1$) is controlled by another mechanism having a higher activation energy of 200-250 kJ/mole. Redfern explained these results on the basis of a process in which the dehydroxylation reaction proceeds inward in crystallites parallel to the (001) planes. The low activation energy at the beginning of the reaction is due to OH^- groups at the crystallite edges escaping without diffusing through the bulk crystallite. The portion of the reaction having a high activation energy comes from the build-up of intracrystalline water vapor pressure as the reaction proceeds. It is important to note, however, that Redfern (1987) failed to obtain confirming physical evidence for his proposed reaction mechanisms nor did he obtain evidence for a P_{H_2O} dependence, although other studies have suggested this dependence.

In contrast to the results on kaolinite by Redfern (1987), Goss (1987) obtained in a study of the transformation of goethite to hematite considerable supplementary physical evidence in support of her proposed mechanisms. She used both isothermal and nonisothermal TGA data collected to about 400°C to obtain information on reaction mechanism and activation energy for this transformation and to derive values of m, the reaction order, from "ln-ln" plots. Her values differed from those obtained by other authors; the differences were ascribed to the considerably longer times for her experiments and to differences in sample characteristics. The "ln-ln" analysis suggested that the early stages of the reaction were controlled by a diffusion mechanism and the later stages by a phase-boundary mechanism. This phase-boundary mechanism was substantiated by a detailed study of heated and unheated goethite in a transmission electron microscope.

Goss (1987) concluded that the reaction proceeded inwards in a planar fashion from the surfaces of crystallites by the formation of parallel pores within the close-packed planes. The reaction product consisted of finely twinned hematite separated by the parallel pores. The reaction front consisted of goethite and the twin-pore product, and the front proceeded primarily parallel to the goethite (100) plane. Apparently, long-range diffusion did not take place.

Activation energies were obtained from isothermal data using both the "change-in-rate" and the "time-to" methods and from nonisothermal data using a Coats and Redfern (1964) plot. Values of activation energy determined using these three methods differed significantly, but for values of α between 0.1 and 0.8, several trends were noted: (1) E increased with decreasing temperature; (2) E increased with decreasing gas flow rate; and (3) E decreased with increasing grain size, the differences being greater at lower temperatures. These differences were attributed to both a change in mechanism with temperature and the fact that the lower-temperature experiments were closer to equilibrium. The equilibrium transformation temperature was calculated to be 120°C.

Inasmuch as the Arrhenius equation applies if the system is far from equilibrium, the activation energy for the highest temperature range represents the transformation most closely. The abrupt decrease in reaction rate at reacted fractions >0.8 was probably due to the prevention of transformation of the central core of goethite due to increased water vapor pressure. The diffusion mechanism indicated at lower temperatures was migration of OH^- or proton-ion transfer across the reaction interface, and the increase in activation energy with decreasing crystallite size suggests that the mechanism was surface-area dependent. Goss (1987) also concluded that the activation energy and the reaction kinetics depended on numerous factors, including temperature, reacted fraction (α), sample size, sample shape, and water vapor pressure.

In a study that is also applicable to mineral dehydration, Ribas *et al.* (1986) examined the multi-stage dehydration of $CoCl_2{\cdot}6H_2O$ using a combination of isothermal and nonisothermal data. $CoCl_2{\cdot}6H_2O$ exhibits four dehydration reactions, although two occur over such a short temperature interval that they could not be separated by TGA. Thus, only three distinct reactions were discernible by TGA or DSC. From their data, Ribas *et al.* (1986) concluded that $CoCl_2{\cdot}6H_2O$ dehydrated through a nucleation-growth mechanism; they also determined the activation energies for the three dehydration reactions. The procedure used by these workers is directly applicable to the study of the dehydration of many smectites and zeolites, for which available TGA data suggest multiple dehydration events. Andersson *et al.* (1983) described a similar isothermal kinetic study of the rehydration of $BaCl_2{\cdot}H_2O$ to $BaCl_2{\cdot}2H_2O$. They conducted experiments at four temperatures and under several water vapor pressures. In addition, they characterized their materials by a combination of XRD, nuclear magnetic resonance, and electron microscopic methods in an effort to understand the factors affecting the hydration kinetics and the effects of repeating the hydration-dehydration reaction many times. Their data show evidence for the presence of capillary condensed water, and they suggested that condensation of this capillary water might have been the rate-controlling step for rehydration. Future studies of mineral hydration-dehydration reactions should consider the procedure used by Andersson *et al.* (1983).

Several important points are illustrated by the experiments of Goss (1987) and others. First, the results demonstrate the importance of determining the effects of varying instrumental and sample parameters on the transformation being studied. These parameters include sample crystallinity, sample size, particle size and packing density, amount of excess structural water, gas flow rate, water vapor pressure, and heating rate. Ninan (1986) showed that sample size can affect kinetic parameters determined for some reactions, whereas other reactions may be completely insensitive to sample size. Using the decomposition of Ca-oxalate monohydrate, Ninan found that the first reaction, involving water loss, was dependent on sample mass. The other thermal reactions of this material were apparently sample independent.

The results of Brindley and Millhollen (1966) and Brindley *et al.* (1967a, 1967b) clearly illustrate the effects of water vapor pressure on dehydroxylation reactions of phyllosilicates. They concluded that kinetic data obtained under water vapor pressures present in normal atmospheric air are of little value, and that the effect of ambient water vapor pressure is not negligible unless pressures are <0.1 torr. In addition, they found that small and loosely packed samples reacted at low heating rates are necessary to maintain this low water vapor pressure within the sample powder. Goss's work (1987) shows that apparently simple reactions may proceed via several steps with different mechanisms and different activation energies and that reaction kinetics must be examined over a wide range of temperature and reacted fraction to determine the possible dependence of mechanism on these factors. Finally, Goss (1987) clearly demonstrated the importance of obtaining additional physical data, such as microstructural information from transmission electron microscopy, to support proposed mechanisms, a procedure that is highly recommended. For example, she showed that the reaction mechanisms elucidated in her study of goethite differed significantly from those obtained in other studies in which narrower ranges of temperature and reacted fraction were investigated.

Modeling dehydration processes in minerals

Although TGA data are quite useful for obtaining information on the kinetics of thermal reactions, they can also be used with available thermodynamic data and solid-solution models to obtain equilibrium information if the processes occurring in the TGA can be shown not to be kinetically limited. The results of TGA experiments, in which mineral dehydration is the process being observed, can be modeled by considering the mineral being dehydrated to be a solid solution of hydrous and anhydrous end members. This approach is reasonable if dehydration does not lead to appreciable modification of the crystal structure, such as in the dehydration of zeolites, and if the dehydration reaction proceeds rapidly and is not kinetically limited.

The dehydration of analcime is a particular reaction for which sufficient data exist to allow modeling and examination of the processes that occur during a TGA experiment. The possible effects of dehydration rate and the question of whether an equilibrium state of dehydration is maintained can be examined. An estimate of P_{H_2O}

can be made, and the equilibrium state of hydration at a P_{H_2O} other than that of the TGA experiment can be estimated. These estimates allow laboratory measurements to be extrapolated to real-life conditions and applications.

The model is based on an ideal solid solution for the analcime-dehydrated analcime system and assumes that vacancies and H_2O mix randomly among the water sites and that no enthalpy, ΔH_{mix}, is associated with the mixing. The entropy associated with mixing can be estimated using the statistical mechanical analog for entropy,

$$S = k \ln \Omega \tag{24}$$

where k is Boltzmann's constant and Ω is the number of distinct ways the hydrous and anhydrous analcime units can be arranged in the mixed crystal. For analcime of composition:

$$NaAlSi_2O_6{\cdot}[N_h/(N_h + N_v)]H_2O.$$

$$\Omega = (N_h + N_v)!/N_h!N_v!, \tag{25}$$

where N_h and N_v are the numbers of water molecules and vacancies present, In unmixed hydrous and anhydrous analcime crystals, the water and vacancies can be distributed only one way. Therefore the entropy of mixing is given by:

$$\Delta S_{mix} = k[\ln(N_h + N_v)! - \ln N_h! - \ln N_v!]]. \tag{26}$$

Using Stirling's approximation ($\ln x! \simeq x \ln x - x$ for very large values of x) and noting that $k = R/N$, where R is the gas constant and N is Avogadro's number,

$$\Delta S_{mix} = -Rn_h\ln[n_h/(n_h + n_v)] + n_v\ln[n_v/(n_h + n_v)], \tag{27}$$

where n_h and n_v are the number of moles of hydrous and anhydrous analcime, respectively. Because

$$\Delta H_{mix} = 0 \tag{28}$$

and

$$\Delta G_{mix} = \Delta H_{mix} - T\Delta S_{mix},$$

the Gibbs free energy of mixing is given by:

$$\Delta G_{mix} = RTn_h\ln[n_h/(n_h + n_v] + n_v\ln[n_v/n_h + n_v)]. \tag{29}$$

From the definition

$$\mu_i = \partial G/\partial n_i \quad {}_{(T,P,n_j,j\neq i)}, \tag{30}$$

where μ_i is the chemical potential of the ith component, it follows that

$$\mu_h - \mu^\circ_h = \partial \Delta G_{mix}/\partial n_{h\,(T,P,nv)}. \tag{31}$$

Differentiating Eq. (29) and combining with Eq. (31) gives

$$\mu_h - \mu^\circ_h = RT \ln x_h, \tag{32}$$

where μ°_h and μ_h are the chemical potential of hydrated analcime in fully hydrated analcime and partially dehydrated analcime, respectively, and $x_h = n_h/(n_h + n_v)$.

The dehydration of analcime in a TGA experiment is represented by the reaction:

hydrated analcime → dehydrated analcime + water.

This reaction will be at equilibrium if

$$\mu_h = \mu_v + \mu_{H_2O} \tag{33}$$

or

$$\mu^\circ_v + \mu^\circ_{H_2O} - \mu^\circ_h = -RT \ln [(x_v f_{H_2O})/f^\circ_{H_2O}]/x_h, \tag{34}$$

where f_{H_2O} is the fugacity of water and $f^\circ_{H_2O}$ is the fugacity at the reference pressure at which $\mu^\circ_{H_2O}$ is defined. Defining the chemical potential of zeolitic water, $u^\circ_{zH_2O}$, by

$$\mu^\circ_{zH_2O} = \mu^\circ_h - \mu^\circ_v \tag{35}$$

gives

$$\mu^\circ_{H_2O} - \mu^\circ_{zH_2O} = -RT \ln[(x_v f_{H_2O})/f^\circ_{H_2O}]/x_h. \tag{36}$$

For a TGA experiment, x_h and $x_v = (1 - x_h)$ can be calculated as a function of temperature from the mass loss data assuming all mass loss is due to the removal of water. If the partial pressure of water is known, Eq. (36) can be used to check the validity of the assumptions of ideal mixing and maintenance of the equilibrium state of dehydration using the data of Johnson *et al.* (1982) for the thermodynamics of hydrated and dehydrated analcime; however, P_{H_2O} is not generally known. In our experiments, dry nitrogen was passed through the TGA apparatus during the experiment giving TGA curves, such as those shown in Figure 23. If the atmosphere with which the samples communicated was actually dry, analcime should have completely dehydrated, even at ambient temperature, given sufficient time. The shape of the mass-loss curve would then have been determined entirely by the rate of dehydration. In support of this possibility, note that analcime reaches a given state of dehydration at higher temperature if the heating rate of the experiment is greater (see Figures 23 and 24).

The dehydration of analcime, however, produces water and must contribute a finite P_{H_2O} to the experiment. Water being driven from the powdered analcime sample must diffuse through the individual crystallites and the pore space in the powder before being carried away by the nitrogen gas stream. The analcime surface may therefore see an appreciable P_{H_2O}, which retards the rate. Eq. (36) can be used to calculate the equilibrium value of P_{H_2O} as a function of temperature and state of hydration. To simplify the calculation, water can be assumed to behave like an ideal gas in these experiments, so that

$$f_{H_2O} = P_{H_2O}. \tag{37}$$

Typical results for three heating rates are shown in Figure 25. The calculated P_{H_2O} is proportional to the rate of mass loss with respect to change in temperature, which is proportional to mass loss per unit time, because constant heating rates were used. Table 1 shows results from experiments on analcime conducted at heating rates of 5°, 10°, and 20°C/min. Maximum calculated P_{H_2O} was proportional to the maximum rate of evolution of H_2O from the sample normalized to the amount of sample present. These results suggest that analcime dehydration in TGA experiments is an equilibrium process and that variation in the mass loss curve with heating rate is caused not by kinetics of dehydration, but by dependence of P_{H_2O} on the heating rate.

Experiments were also performed on analcime in which the amount of sample and the gas flow rate were varied while maintaining the heating rate at 10°C/min. The results of these experiments are shown in Table 2 and Figure 26. Most experiments were made at a gas flow rate of 50 cm³/min. The experiment made with no gas flow showed a small, perhaps insignificant, decrease in the maximum rate of mass loss and a significant increase in calculated maximum P_{H_2O}. An increase in P_{H_2O} seems reasonable due to a build-up of water vapor in the apparatus. The slightly slower dehydration rate was probably due to slow vapor transport in the absence of a flowing gas. Increasing the sample size had no significant effect on the maximum sample-mass-normalized rate of mass loss or on the calculated maximum P_{H_2O}, probably because of the increased pore volume of the sample. Although more water was being driven from the sample, it was being driven into a proportionately larger pore volume, with the result that P_{H_2O} remained about the same.

The results presented here suggest that the thermal decomposition of analcime is an equilibrium process and that TGA experiments can yield useful information pertaining to the hydration of this mineral. The present TGA data, however, are of limited value in this respect, because P_{H_2O} in the experiments is unknown. This is because the P_{H_2O} of concern was that present in the pore volume of the sample and not that in the bulk of the gas in the thermobalance. The current belief is that P_{H_2O} in the pore volume can be known only if the gas in the apparatus is pure water vapor. Under any other circumstances, the potential exists for a gradient in P_{H_2O} from the pore volume of the sample to the open volume of the thermobalance. Such an experimental configuration has the advantage that the equilibrium nature of the experiment

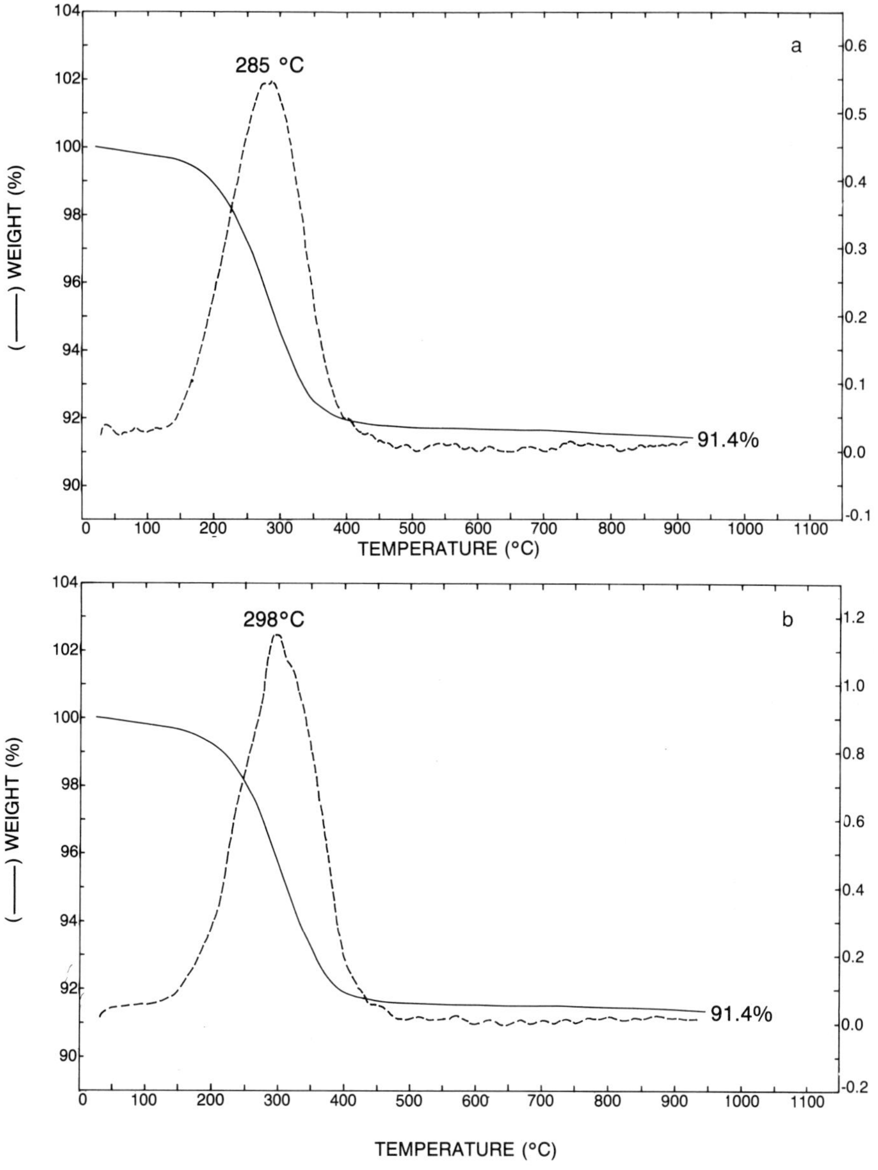

Figure 23. Thermogravimetric analysis curves for analcime (Iceland) showing shift in inflection temperature from (a) 285°C (10°C/min heating rate, 14.24-mg sample) to (b) 298°C (20°C/min heating rate, 16.32-mg sample) with an increase in heating rate (50-cm^3/min N_2 purge).

could be verified by reversing the experiment through a complete heating and cooling cycle. If such experiments are performed, $\mu°_{zH_2O}$ could be calculated as a function of temperature, making the calculation of the equilibrium state of hydration of analcime possible at any temperature and P_{H_2O}.

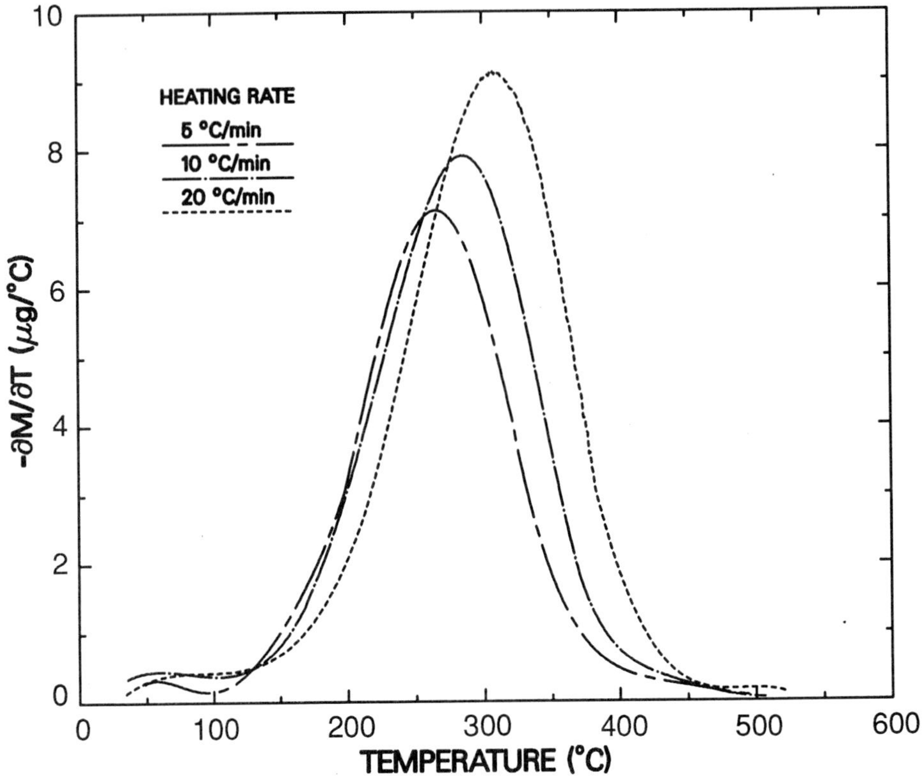

Figure 24. Differential thermogravimetric (DTG) curves for analcime (Iceland) heated at three rates (5°C/min, 12.30-mg sample; 10°C/min 14.24-mg sample; 20°C/min, 16.32-mg sample) showing increase in temperature of DTG peak with an increase in heating rate (50-cm^3/min N_2 purge). Heights of DTG peaks are proportional to sample mass.

So far, only the dehydration of analcime has been discussed, but this method probably can be extended to any material for which removal of material during TGA leaves the remaining structure intact. Solid solution theory cannot be applied to structures that are altered by the removal of water or some other easily mobilized component, because vacancies cannot be assumed to exist at the sites of the removed material. This treatment, therefore, should not be applied to TGA experiments on materials in which the crystal structure of the material is altered. Most zeolites probably fall into the class of materials that can be treated in this way, but many may have multiple water sites. If these sites are equivalent, Eq. (32) becomes:

$$\mu_h - \mu^\circ_h = \alpha RT \ln x_h. \qquad (38)$$

where α is the number of independent sites. If the sites are inequivalent (as for clinoptilolite, in which the derivative mass-loss curve shows multiple peaks (see Figures 17 and 18)] and are independent (i.e., the way one set of sites is occupied does not affect another set), Eq. (36) can be used for each set of sites independently.

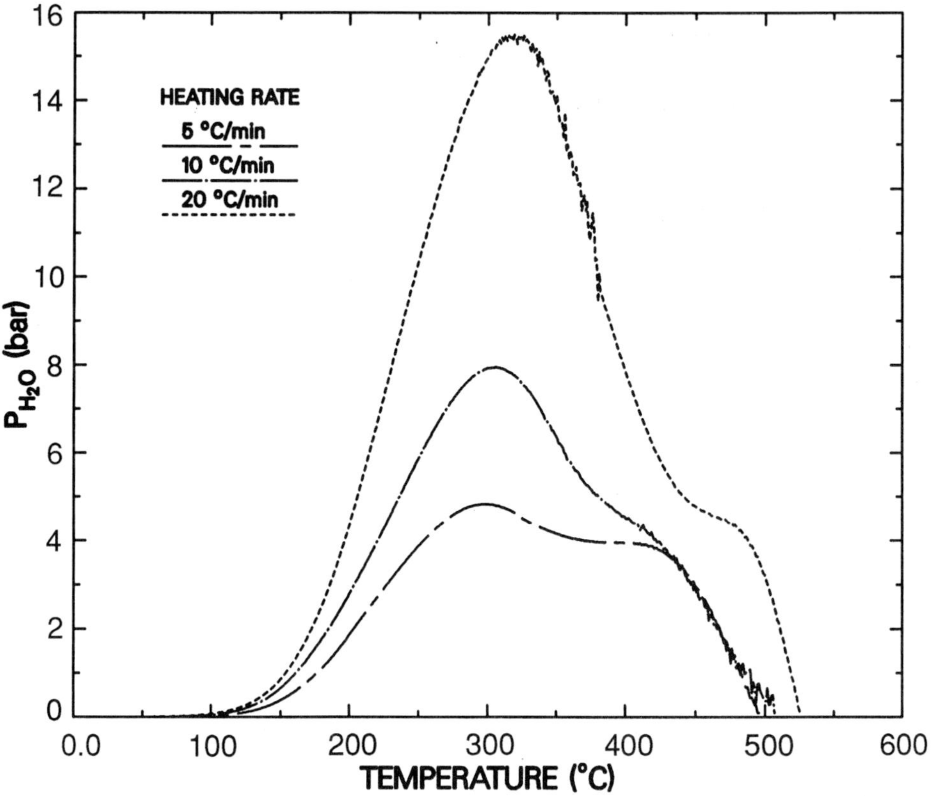

Figure 25. Calculated water vapor pressure, P_{H2O}, for analcime (Iceland) heated at three rates (5°, 10°, 20°/min; sample masses as in Figure 24, showing an increase in calculated P_{H2O} with heating rate (50-cm^3/min N_2 purge).

Table 1. Maximum rate of mass loss and maximum calculated P_{H_2O} for thermogravimetric analysis experiments on analcime at different heating rates.

Heating rate (°C/min)	(1) Max (∂mass/∂time) (μg/mg·min)	(2) Max P_{H_2O} (bar)	(2)/(1) (bar·mg· min/μg)
5	2.95	0.048	0.016
10	5.50	0.079	0.014
20	11.20	0.115	0.014

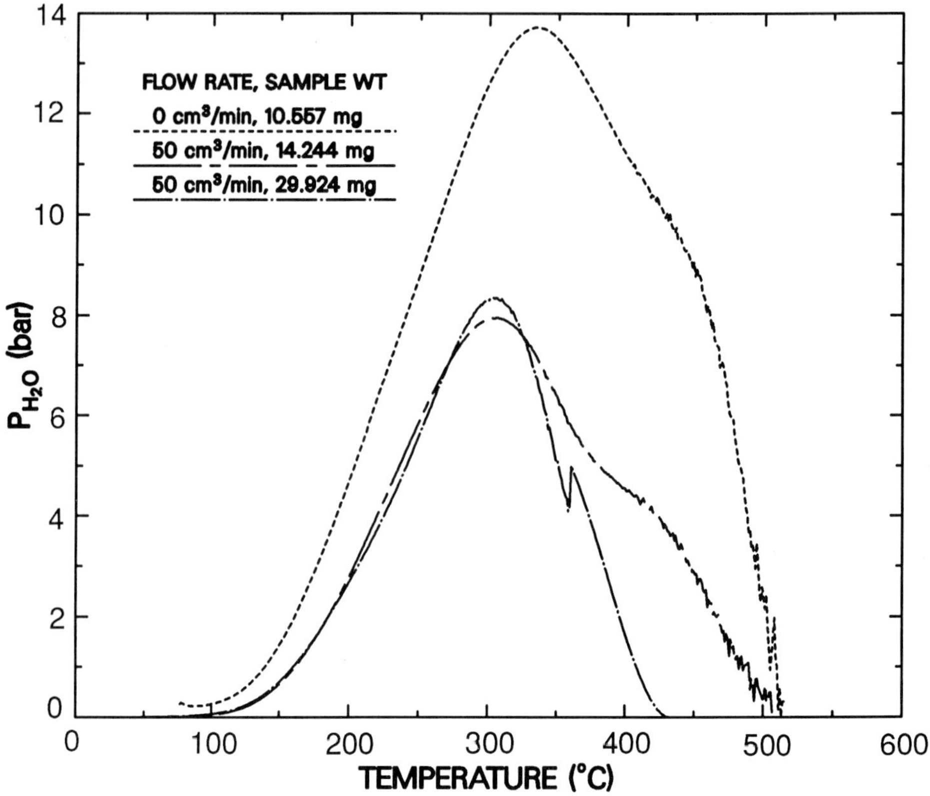

Figure 26. Calculated water vapor pressure, P_{H_2O}, for analcime (Iceland) samples of different mass and with different flow rates showing the significant effect of purge-gas flow rate and lack of dependence of P_{H_2O} on sample mass (10°C/min heating rate for all samples; sample mass and N_2 purge gas rate shown in figure).

Table 2. Maximum rate of mass loss and maximum calculated P_{H_2O} for thermogravimetric analysis experiment on analcime at different gas flow rates.

Gas flow (cm³/min)	(1) Max (∂mass/∂time) (μg/mg·min)	(2) Max P_{H_2O} (bar)	(2)/(1) (bar·mg·min/μg)
50	5.50	0.079	0.014
0	5.00	0.137	0.027
50	5.40	0.084	0.016

A different $\mu°_{zH_2O}$ would then occur for H_2O in each distinct set of sites. For a given f_{H_2O}, the state of hydration on each independent type of water site, x_v and x_h would be determined using Eq. (36) and the appropriate $\mu°_{zH_2O}$. The overall state of hydration at a particular temperature would be determined by summing the states of hydration of the independent types of sites, taking into account the proportion of the total water present at each of the types of sites. If the different types of sites are not independent, knowledge of the interaction between sites would be necessary before detailed modeling could be done.

SUMMARY AND CONCLUSIONS

Thermogravimetric analysis is a powerful method capable of providing quantitative information on minerals that is difficult to obtain with other techniques. The method can rapidly provide insights into materials and can be used as a quick identification tool. For example, the structure type and order/disorder of kaolin-group minerals can be thoroughly characterized using TGA In addition, TGA can provide important quantitative data on mixtures, data that for some samples would prove difficult to obtain by other means. Although data such as these can be quickly and easily obtained, the strengths of TGA lie in its use in elucidating the nature of water and other volatiles in minerals. For example, the effects of interlayer-cation or exchangeable-cation composition on the thermal behavior of clays and zeolites can be thoroughly characterized, using the results as a predictive tool. The results of such studies can also provide information as to the nature of the sites occupied by interlayer or exchangeable cations, information that is complementary to that obtained by other methods, such as X-ray powder diffraction.

Perhaps most importantly for applied studies, TGA can provide considerable information on the effects of variables, such as temperature and water vapor pressure, on thermal reactions and on the kinetics of any thermal reaction involving a mass loss or gain. Kinetic studies and equilibrium modeling, coupled with variation of appropriate parameters, can provide valuable predictive data that can be used in many environmental and industrial settings.

ACKNOWLEDGMENTS

The authors are grateful to W. H. Casey, Sandia National Laboratory, for discussions on reaction kinetics and for comments on the manuscript, to M. A. Carpenter, University of Cambridge, for preprints of the Redfern and Goss papers, and to S. J. Chipera, D. Horton, G. D. Guthrie, D. T. Vaniman, and an anonymous reviewer for valuable comments on the manuscript.

REFERENCES CITED

Andersson, J., Azoulay, M., and DePablo, J. (1983) Kinetic investigation of the sorption of water by barium chloride monohydrate: *Thermochim. Acta* **70,** 291-302.

Araya, A. and Dyer, A. (1981) Studies on natural clinoptilolites: *J. Inorg. Nucl. Chem.* **43,** 589-594.

Arnold, M., Veress, G. E., Paulik, J., and Paulik, F. (1979) Problems of the characterization of thermoanalytical processes by kinetic parameters: *J. Therm. Analy.* **17,** 507-528.

Bagchi, T. P. and Sen, P. K. (1983) Redundancy of isothermal kinetic data to back up nonisothermal kinetic data: *Thermochim. Acta* **70,** 363-365.

Barrow, G. M. (1973) *Physical Chemistry,* McGraw-Hill, New York, p. 646.

Bassett, W. A. (1960) Role of hydroxyl orientation in mica alteration: *Geol. Soc. Amer. Bull.* **71,** 449-456.

Bish, D. L (1988) Effects of composition on the dehydration behavior of clinoptilolite and heulandite: in *Occurrence, Properties, and Utilization of Natural Zeolites,* D. Kallo and H. S. Sherry, eds., Akademiai Kiado, Budapest, 565-576.

Bish, D. L. and Brindley, G. W.(1977) A reinvestigation of takovite, a nickel aluminum hydroxy-carbonate of the pyroaurite group: *Amer. Mineral.* **62,** 458-464.

Brindley, G. W. (1961) The role of crystal structure in the dehydration reactions of some layer-type minerals: *J. Mineral. Soc. Japan* **5,** 217-237.

Brindley, G. W. and Millhollen, G. L. (1966) Chemisorption of water at high temperatures on kaolinite: Effect on dehydroxylation: *Science* **152,** 1385-1386.

Brindley, G. W. and Nakahira, M. (1959) X-ray diffraction and gravimetric study of the dehydration reactions of gibbsite: *Z. Kristallogr.* **112,** 136-149.

Brindley, G. W, Sharp, J. H., Patterson, J. H., and Achar, B. N. N. (1967a) Kinetics and mechanism of dehydroxylation processes: I. Temperature and vapor pressure dependence of dehydroxylation of kaolinite: *Amer. Mineral.* **52,** 201-211.

Brindley, G. W., Sharp, J. H., Patterson, J. H., and Achar, B. N. N. (1967b) Kinetics and mechanism of dehydroxylation processes: II. Temperature and vapor pressure dependence of dehydroxylation of serpentine: *Amer. Mineral.* **52,** 1697-1705.

Cabrera, J. G. and Eddleston, M. (1983) Kinetics of dehydroxylation and evaluation of the crystallinity of kaolinite: *Thermochim. Acta* **70,** 237-247.

Calef, D. F. and Deutsch, J. M. (1983) Diffusion-controlled reactions: *Ann. Rev. Phys. Chem.* **34,** 493-517.

Carroll, B. and Manche, E. P. (1972) Kinetic analysis of chemical reactions for nonisothermal procedures: *Thermochim. Acta* **3,** 449-459.

Coats, A. W. and Redfern, J. P. (1963) Thermogravimetric analysis. A review: *Analyst* **88,** 906-924.

Coats, A. W. and Redfern, J. P. (1964) Kinetic parameters from thermogravimetric data: *Nature* **201,** 66-69.

Crank, J. (1975) *The Mathematics of Diffusion:* Oxford Science Publ., London, 414 pp.

Dondur, V. and Vucelic, D. (1983a) An approach to the kinetics of water desorption from A-zeolites. Part I. Isothermal and nonisothermal desorption: *Thermochim. Acta* **68,** 91-99

Dondur, V. and Vucelic, D. (1983b) An approach to the kinetics of water desorption from A-zeolites. Part II. Resolution of the complex dehydration process to elementary reactions: *Thermochim. Acta* **68,** 101-111.

Dondur, V. and Vucelic, D. (1983c) An approach to the kinetics of water desorption from A-zeolites. Part III. Thermodesorption of complexes with nonuniform activation energies: *Thermochim. Acta* **68,** 113-119.

Doyle, C. D. (1966) Quantitative calculations in thermogravimetric analysis: in *Techniques and Methods of Polymer Evaluation*, P. E. Slade, Jr. and L. T. Jenkins, eds., Marcel Dekker, New York, 113-216.

Earnest, C. M. (1980) The application of differential thermal analysis and thermogravimetry to the study of kaolinite clay minerals: *Perkin-Elmer Thermal Analysis Application Study* **30**, 11 pp.

El-Akkad, T. M., Flex, N. S., Guindy, N. M., El-Massry, S. R., and Nashed, S. (1982) Thermal analyses of mono- and divalent montmorillonite cationic derivatives: *Thermochim. Acta* **59,** 9-17.

El-Barawy, K. A., Girgis, B. S., and Felix, N. S. (1986) Thermal treatment of some pure smectites: *Thermochim. Acta* **98,** 181-189.

Flynn, J. H. and Dunlap, L. A. (1986) Temperature gradients in horizontal tube furnaces: *Thermochim. Acta* **105,** 215-218.

Galli, E., Gottardi, G., Mayer, H., Preisinger, A., and Passaglia, E. (1983) The structure of potassium-exchanged heulandite at 293, 373 and 593 K: *Acta Crystallogr.* **B39,** 189-197.

Giese, R. F., Jr. (1971) Hydroxyl orientation in muscovite as indicated by electrostatic energy calculations: *Science* **172,** 263-264.

Giese, R. F., Jr. (1978) The electrostatic interlayer forces of layer structure minerals: *Clays & Clay Minerals* **26,** 51-57.

Giese, R. F., Jr. (1979) Hydroxyl orientations in 2:1 phyllosilicates: *Clays & Clay Minerals* **27,** 213-223.

Goss, C. J. (1987) The kinetics and reaction mechanism of the goethite to hematite transformation: *Mineral. Mag.* **51,** 437-452.

Grim, R, E.(1953) *Clay Mineralogy:* McGraw-Hill, New York, 384 pp.

Grim, R. E. and Kulbicki, G. (1961) Montmorillonite: High temperature reactions and classification: *Amer. Mineral.* **46,** 1329-1369.

Hancock, J. D. and Sharp, J. H. (1972) Method of comparing solid-state kinetic data and its application to the decomposition of kaolinite, brucite, and $BaCO_3$: *J. Amer. Ceram. Soc.* **55,** 74-77.

Johnson, G. K., Flotow, H. E., O'Hare, P. A. G., and Wise, W. S. (1982) Thermodynamic studies of zeolites: Analcime and dehydrated analcime: *Amer. Mineral.* **67,** 736-748.

Johnson, L. J., Chu, C. H., and Hussey, G. A. (1985) Quantitative clay mineral analysis using simultaneous linear equations: *Clays & Clay Minerals* **33,** 107-117.

Keattch, C. J. and Dollimore, D. (1975) *An Introduction to Thermogravimetry:* Heyden, London, 164 pp.

Knowlton, G. D., White, T. R., and McKague, H. L. (1981) Thermal study of types of water associated with clinoptilolite: *Clays & Clay Minerals* **29,** 403-411.

Koster van Groos, A. F. and Guggenheim, S. (1984) The effect of pressure on the dehydration reaction of interlayer water in Na-montmorillonite (SWy-1): *Amer. Mineral.* **69,** 872-879.

Koster van Groos, A. F. and Guggenheim, S. (1986) Dehydration of K-exchanged montmorillonite at elevated temperatures and pressures: *Clays & Clay Minerals* **34,** 281-286.

Koster van Groos, A F. and Guggenheim, S. (1987) The dehydration of Ca- and Mg-exchanged montmorillonite (SWy-1) at elevated pressures: *Amer. Mineral.* **72,** 292-298.

Koster van Groos, A. F. and Guggenheim, S. (1990) High-pressure differential thermal analysis: Applications to clay minerals: in *CMS Workshop Lectures, Vol. 3, Thermal Analysis in Clay Science,* J. W. Stucki, D. L. Bish, and F. A. Mumpton, eds., The Clay Minerals Society, Boulder, Colorado, 49-94.

Lasaga, A. (1986) Metamorphic reaction rate laws and development of isograds: *Mineral. Mag.* **50,** 359-373.

MacKenzie, K. J. D. and Bowden, M. E. (1983) Thermal and Mössbauer studies of iron-containing hydrous silicates. IV. Amesite: *Thermochim. Acta* **64,** 83-106.

MacKenzie, K. J. D., Berezowski, R. M., and Bowden, M. E. (1986) Thermal and Mössbauer studies of iron-containing hydrous silicates. VI. Minnesotaite: *Thermochim. Acta* **99,** 273-289

Mumpton, F. A. (1960) Clinoptilolite redefined: *Amer. Mineral.* **45,** 351-369.

Murray, P. and White, J. (1949) Kinetics of the thermal dehydration of clays: *Trans. Brit. Ceram. Soc.* **48,** 187-206.

Nernst, W. and Riesenfeld, E. H. (1903) Über quantitative Gewichtanalyse mit sehr kleinen Substanzmengen: *Chem. Berichte* **36,** 2086-2093.

Ninan, K. N. (1986) Thermal decomposition kinetics. Part XIV. Kinetics and mechanism of isothermal decomposition of calcium oxalate monohydrate and correlations with sample mass: *Thermochim. Acta* **98,** 221-236.

Nutting, P. G. (1943) Some standard thermal dehydration curves of minerals: *U.S. Geol. Surv. Prof. Pap.* **197-E,** 197-217.

Onike, F., Martin, G. D., and Dunham, A. C. (1986) Time-temperature-transformation curves for kaolinite: *Materials Sci. Forum* **7,** 73-82.

Putnis, A. and McConnell, J. D. C. (1980) *Principles of Mineral Behaviour:* Elsevier, New York, 257 pp.

Redfern, S. A. T. (1987) The kinetics of dehydroxylation of kaolinite: *Clay Mineral.* **22,** 447-456.

Ribas, J., Escuer, A., Serra, M., and Vicente, R (1986.) Solid state kinetic parameters of the dehydration of $CoCl_2{\cdot}6H_2O$ by isothermal thermogravimetric measurements: *Thermochim. Acta* **102,** 125-135.

Rosen, J. B. (1952) Kinetics of a fixed bed system for solid diffusion into spherical particles: *J. Chem. Phys.* **20,** 387-394.

Ross, C. S. and Hendricks, S. B. (1945) Minerals of the montmorillonite group. Their origin and relation to soils and clays: *U.S. Geol. Surv. Prof. Pap.* **205B,** 79 pp.

Satava, V. (1971) Mechanism and kinetics from nonisothermal TG traces: *Thermochim. Acta* **2,** 423-428.

Sestak, J. (1966) Errors of kinetic data obtained from thermogravimetric curves at increasing temperature: *Talanta* **13,** 567-579.

Shannon, R. D. (1976) Revised effective ionic radii in halides and chalcogenides: *Acta Crystallogr.* **A32,** 751-767.

Simonot-Grange, M. H.(1979) Thermodynamic and structural features of water sorption in zeolites: *Clays & Clay Minerals* **27,** 423-428.

Smykatz-Kloss, W. (1974) DTA as a tool for the measurement of disorder in kaolinites and for the classification of montmorillonites: in *Thermal Analysis, Proc. 4th Int. Conf. Thermal Analysis, Vol. 2,* 561-565.

van Olphen, H. and Fripiat, J. J. (1979) *Data Handbook for Clay Materials and other Non-Metallic Minerals:* Pergamon Press, Oxford, United Kingdom, 346 pp.

Van Reeuwijk, L. P. (1974) The thermal dehydration of natural zeolites: *Meded. Landbouwhogeschool Wageningen* **74,** 1-88

Weaver, C. E. and Pollard, L. D. (1973) *The Chemistry of Clay Minerals:* Elsevier, Amsterdam, 213 pp.

Wendlandt, W. W. (1986) *Thermal Analysis:* Wiley, New York, 814 pp.

VACUUM THERMOGRAVIMETRIC ANALYSIS AND EVOLVED GAS ANALYSIS BY MASS SPECTROSCOPY

F. J. Wicks and R. A. Ramik

CONTENTS

Introduction	160
Instrumentation	162
Thermal balance	162
Mass spectrometer	162
Vacuum system	164
The Royal Ontario Museum system	164
Methodology	167
Sample preparation	167
Sample storage	169
Sample weighing	169
Vacuum thermogravimetric analysis	170
Evolved gas analysis	171
Interpretation	172
Summary	182
Advantages of the vacuum method	182
Disadvantages of the vacuum method	183
Conclusions	184
Acknowledgments	185
References Cited	185

VACUUM THERMOGRAVIMETRIC ANALYSIS AND EVOLVED GAS ANALYSIS BY MASS SPECTROSCOPY

F. J. WICKS AND R. A. RAMIK

Department of Mineralogy
Royal Ontario Museum
Toronto, Ontario M5S 2C6
Canada

INTRODUCTION

The combined techniques of vacuum thermogravimetric analysis (TGA) and evolved gas analysis (EGA) by mass spectrometry (MS) can be productive tools for mineral studies. In its simplest configuration, a TGA-EGA system correlates quantitative thermal weight-loss with semiquantitative evolved gas data. Thus, each weight-loss event can be assigned to the volatile species identified in that temperature interval. More sophisticated systems are capable of quantitative EGA data, with volatile abundances determined directly by integration of evolved gas peaks and verified with TGA measurements. Alternatively, the ratio of evolved gases may be used to apportion weight losses.

The vacuum TGA-EGA results from cyanotrichite, $Cu_4Al_2(SO_4)(OH)_{12}{\cdot}2H_2O$, illustrate the importance and power of the combined technique. Even the small size sample (1.18 mg) used in this experiment produces good thermogravimetric information (Figure 1) because the losses during heating are relatively large. The TGA curve consists of five distinct weight-loss steps plus minor inflections that suggest further complexities. Except for assuming that H_2O will be lost at a lower temperature than OH, such a curve is impossible to interpret with any certainty without further information. The EGA curves (Figure 1) provide this information. The curves of ionized water[1], H_2O^+, and one of its fragments, O^+, which peak at 61°C, indicate that the first weight-loss event is due to the loss of H_2O. Similarly, the second two events, which peak at 168°C and 308°C, represent the evolution of H_2O during dehydroxylation of the mineral. The sensitivity of the instrument is demonstrated by the detection of the onset of sulfate loss at 300°C where SO_2 is first detected. The major sulfate loss, however, is at 612°C, coincident with the fourth major step on the TGA curve. The fifth and final step is due to the reduction of Cu^{2+} to Cu^+ with a loss of O_2 (O^+ fragment recorded), the only significant evolved gas present in this part of the analysis. Thus, the evolved gas associated with each weight loss can be identified, and the nature of the thermal disassociation is more clearly understood.

[1]The quadrupole mass spectrometer ionizes evolved gas molecules in order to detect them and to measure their apparent masses. To some extent, the molecules are also broken into ionized fragments.

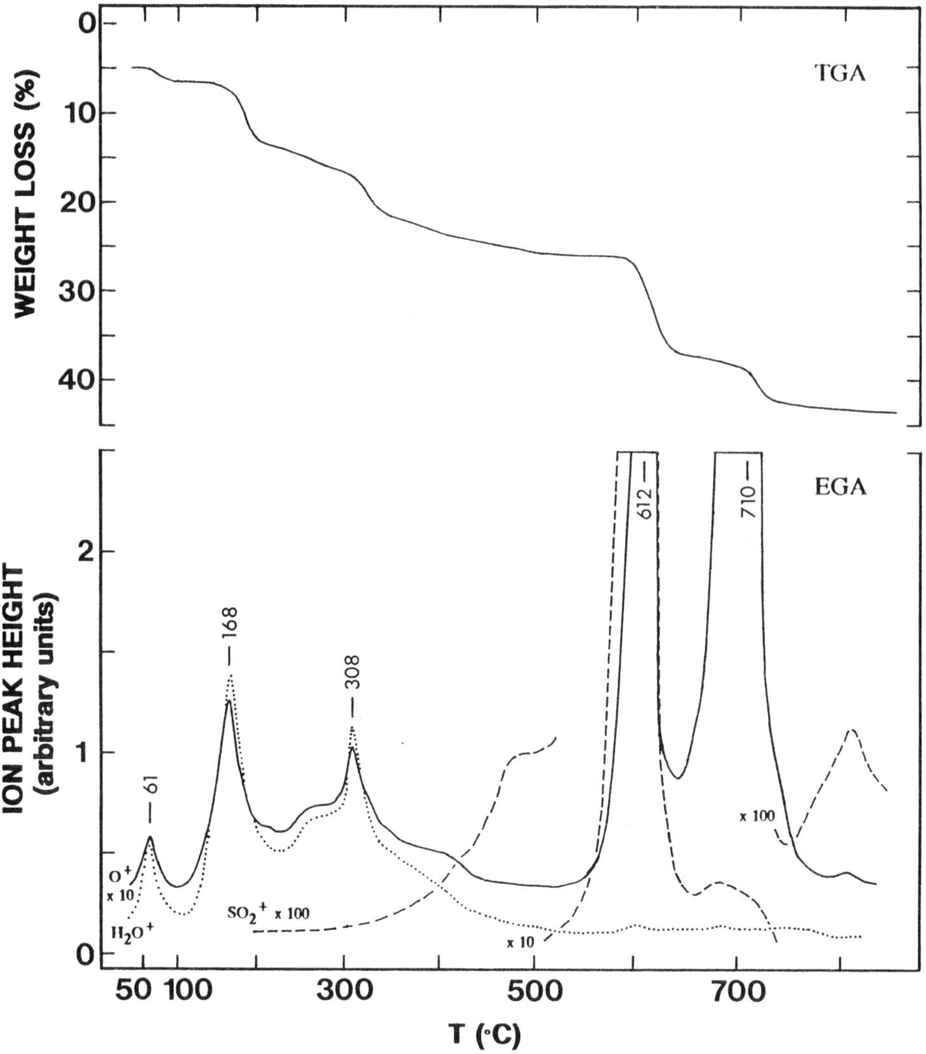

Figure 1. Thermal gravimetric-evolved gas analysis curves of cyanotrichite (1.18 mg) Cap Garonne, France, ROM #M7086. Sample heated at 10°C/min in an initial vacuum of 1 x 10^{-7} torr.

The form and content of this chapter are strongly influenced by our experiences in the thermal analytical laboratory at the Royal Ontario Museum (ROM). The thermal analytical laboratory is directed primarily to the analysis of the volatiles content of minerals by TGA and EGA. Several mineral groups have been examined, among them carbonates, hydroxides, sulfates, borates, phosphates, arsenates, and silicates, including clay minerals and zeolites. Numerous newly discovered species have been analyzed as a part of initial definitive studies. The technique has also been applied to such varied materials as natural glass, metal hydride, fluorinated carbon, and synthetic zeolites. A few rocks have been analyzed with limited success.

The details of coupling a mass spectrometer to a thermoanalyzer were described by Clinckemaille and Hofmann (1972), Szekely and Till (1974), and Dunner and Eppler (1975). Wendtlandt (1986) recently reviewed thermobalance EGA instrumentation and TGA-EGA techniques and applications. Dollimore *et al.* (1984), Holdiness (1984), and Kaiserberger and Emmerich (1982) reviewed and specifically discussed EGA by mass spectrometry, and Morgan *et al.* (1988) and Gallagher (1982) reviewed applications of EGA to earth science and inorganic material studies, respectively. Other relevant papers on high-temperature topics, such as MS analysis of molecular beams, effusion cells, and high-pressure sampling techniques using skimmers, can be found in Hastie (1979).

INSTRUMENTATION

Thermal balance

The thermal balance is the central component of a TGA-EGA system. Each thermal balance on the market has its own particular configuration with specific advantages and disadvantages. The choice of instrument depends on the application and, if necessary, on how the balance must be modified to accommodate the desired EGA instrumentation. Many instruments are compactly constructed and are designed specifically for atmospheric TGA; few are designed for high-vacuum studies. If neither quantitative EGA nor measurement of readily condensable volatiles, such as alkalies, halides, or various metals, are of interest, the gas analyzer need not be in line-of-sight with, nor in close proximity to the sample, and modification may be fairly simple. More demanding applications, such as dynamic, high-temperature analysis of trace volatiles in a milligram-size sample, require an instrument designed with sub-microgram sensitivity, an ultimate vacuum better than 10^{-7} torr, a balance mechanism unaffected by pressure changes, a short line-of-sight path from the sample surface to the detector, a furnace chamber of small volume, and good heating control to 1400°C. In practice, major difficulties, such as the alignment of the detector with the sample, are likely to be encountered in the modification of an off-the-shelf instrument. If, in addition, the same instrument also must be capable of operating at atmospheric pressure and performing differential thermal analysis (DTA), further problems will arise. The difficulty in modifying an instrument increases with the addition of increased capabilities, and some problems may be best solved by the use of more than one instrument. For example, a conventional vacuum TGA-DTA instrument might be used in conjunction with a Knudson cell-mass spectrometer system.

Mass spectrometer

Several gas analysis techniques are available for thermal analysis: mass spectrometry (MS), Fourier-transform infrared (FTIR) spectrometry and other infrared methods (Low, 1967), gas chromatography (GC), manometry (Harris, 1981), and an assortment of specific gas detectors. GC and FTIR spectrometry are increasingly used alone and in combination with MS. Prime (1987) has compared the merits and limitations of MS and IR in conjunction with TGA; however, for vacuum analysis the choice is practically restricted to MS.

Several mass spectrometers are available to match specific analytical problems. The dynamic nature of TGA-EGA requires an analyzer capable of rapid gas analysis with moderate precision, sensitivity, and stability. The quadrupole mass spectrometer with its rapid response best meets these requirements. Some analyses, involving high mass numbers or very complex decompositions, may require the greater resolution provided by magnetic-sector mass spectrometers. These are more expensive and generally slower, thus sacrificing rapid response. Some single-focusing magnetic mass spectrometers are capable of resolving ion doublets and triplets, such as mass 14 (N, CH_2), 16 (O, NH_2, CH_4), 28 (CO, N_2), and 32 (O_2, S), which are most likely to be encountered in mineral analyses (Elliot, 1969). This resolution is beyond the capabilities of quadrupole mass spectrometers, although the problem can be addressed, but with less precision, by the analysis of cracking patterns (discussed below).

The need for high sensitivity can be best met by placing the spectrometer ion source in line with the specimen surface, using a suitable beam skimmer, and using an ion source designed specifically for molecular beam analysis. The low-end magnetic-sector mass spectrometers are cost competitive with the moderate-end to high-end quadrupole instruments (Rosen *et al.*, 1988), but quadrupole mass spectrometers provide a good balance of fast response to the dynamic heating of the system, a reasonable analytical range of masses, and good sensitivity at a reasonable cost.

The theory of quadrupole mass spectrometry was discussed by Dawson (1976). Basic quadrupole spectrometers operate by bombarding a gas at low pressure (generally $<10^{-4}$ torr) with electrons of sufficient energy to allow molecular ionization. In the process, a certain proportion of the molecules is broken into ionized fragments. Positively charged fragments are usually used, but some instruments can also use negatively charged fragments. The charged molecules and fragments are accelerated and then passed through a filter of radio frequency (RF) and direct current (DC) fields toward a detector. For a given RF-DC combination, only ions of a specific mass/charge (m/e) ratio will successfully oscillate through the filter to the detector. The instrument rapidly steps through all necessary combinations to produce a mass spectrum over a given range of m/e. The mass values and peak intensities for the parent molecule and all fragments of the parent molecule are referred to as the cracking pattern for the gas. The mass spectrum is the sum of cracking patterns of all gases present. Some caution must be used in interpreting mass spectra because, as shown by Breth *et al.* (1983), cracking patterns of mixtures vary with pressure.

For most applications, the minimum specifications of a quadrupole spectrometer include a mass range of 1-150 m/e, a capability for rapid scans (several/min) over the entire range, and a quasi-continuous monitoring of at least 10 selected masses. For complex decompositions, the monitoring of more than 15 masses may be required. An electron multiplier is required for the detection of trace quantities of volatiles.

Vacuum system

The type of vacuum pump and the pumping capacity required for a vacuum system depend on the sample size, the volatile content of the sample, and the type of analysis being carried out. The following factors must be considered:

1. The pumping speed must be sufficient to maintain an adequately high vacuum under the load of gases produced during analysis.

2. All gases must be pumped efficiently. Diffusion and turbomolecular pumps do not pump H_2 and, to a lesser degree, He, as efficiently as heavier gases. If H_2 is of particular interest, a turbomolecular pump with the highest possible H_2 compression ratio must be used to obtain as low a H_2 background as possible. If diffusion pumps are selected, fitting them with cold traps is essential to assist in pumping H_2O, which is typically the most abundant residual gas and commonly the principal evolved gas, and to eliminate backstreaming of vacuum pump vapor into the system.

3. The pumping system must operate smoothly. If diffusion pumps are used, the pumping fluid charges may have to be reduced to the minimum to achieve smooth pumping. This will decrease noise in the 0.1-1 Hz range in evolved gas curves.

4. The contaminants in the system must be controlled. Diffusion pump fluids should be selected so that their presence or their bakeout products in the vacuum system are not deleterious to spectrometer performance.

Theoretical and practical high vacuum information can be found in Green (1968), Robinson (1968), and Van Atta (1965).

The Royal Ontario Museum system

The thermal analysis system at the Royal Ontario Museum (ROM) is a good example of the sort of modifications that are required to achieve simultaneous TGA and EGA. Inasmuch as our practical experience is limited to this particular instrument we have chosen to use it as the main example for this chapter. The equipment typifies an old generation of thermal analyzer (Mettler TA-1 Thermoanalyzer *circa* 1972) that has been upgraded by the addition of a modest quadrupole mass spectrometer (Inficon IQ200) (Figure 2). Additional upgrading using two personal computers,

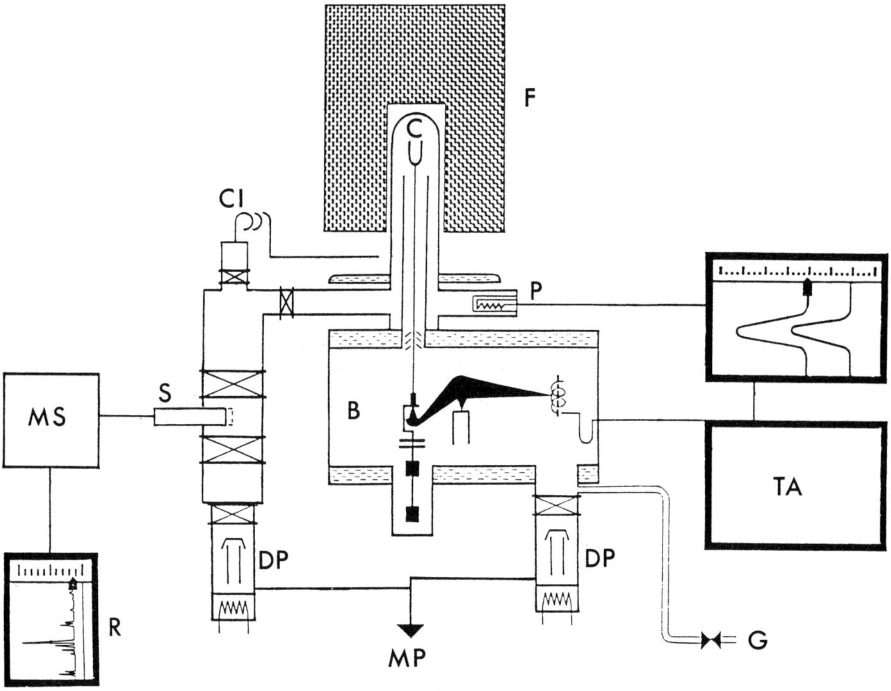

Figure 2. Schematic drawing of the thermoanalyzer-quadrupole mass spectrometer system used at the Royal Ontario Museum: B = balance chamber, C = crucible, CI = capillary inlet system, DP = diffusion pump, F = furnace, G = gas supply, MP = mechanical vacuum pump, MS = quadrupole mass spectrometer controls, P = Penning gauge, R = recorder, S = spectrometer sensor, TA = thermoanalyzer controls (modified from Mettler instruction manual).

one for control of the instrument, and one for data acquisition and data processing, is under way. The Mettler instrument is capable of simultaneous TGA-DTG-DTA (DTG represents derivative themogravimetry, the first derivative of the TGA curve is the rate of weight-loss) in a variety of atmospheres, including corrosive gases and water vapor, as well as high vacuum. Many modifications of the instrument are possible due to the open architecture of the frame. The vacuum system consists of two diffusion pumps (each 135 liter/s) using Santovac 5 fluid and fitted with liquid N_2 cold traps. One pump evacuates the balance chamber, and the other evacuates the furnace via the mass spectrometer line. The original Viton (an elastomeric fluorocarbon) O-rings in the system have been replaced with copper seals wherever possible. A crucial improvement in vacuum-tightness was achieved by replacing the standard Teflon seal at the base of the removable furnace chamber with a seal cut from flawless 1/32" sheet Viton. These modifications resulted in a vacuum of better than 1 x 10^{-7} torr at the sample and about 10^{-8} torr at the spectrometer sensor. The mass spectrometer is thus able to sample gases over a large dynamic range (10^{-4} to 10^{-8} torr) without the need for pressure reduction. A heated capillary system is also available for transferring gases from a 1000°C furnace at atmospheric pressure into the vacuum line of the mass spectrometer. A six-channel recorder is currently used to plot TGA, amplified (x 10) TGA, DTA, DTG, T, and a total pressure signal from a Penning gauge located at the base of the furnace (Figure 2).

Although the thermoanalyzer accommodates a sample as large as a few grams, the maximum permissible sample size for analysis in vacuum is determined by the volatile content of the sample, the heating rate, and the pumping speed of the vacuum system. In the ROM system for example, a 10-mg sample containing 5 wt. % volatiles, heated at 10°C/min, pumped with the two diffusion pumps of 135 liter/s each, typically gives very strong EGA peaks with little likelihood of vacuum overload. The TGA curve on the strip chart is readable to 5 μg and is stable to within 10 μg during analysis, which will be improved by computer control of the system.

Simultaneous DTA is possible on this instrument, but it is not normally used in high-vacuum work because the baseline is poor due to the difficulty in controlling the heating rate at temperatures <300°C. The release of adsorbed gases by a reference material during the analysis of very small samples is an additional concern.

A Mettler 1600°C furnace is used for high-temperature studies; however, analyses in high vacuum are limited to temperatures <1450°C, because at higher temperature, the pressure of O_2 and CO increases rapidly, the former presumably being produced by dissociation of alumina in the furnace lining. Crucibles of Pt-10%Rh are preferred to those of sintered alumina because they are generally easier to clean, less permeable to melts, and less adsorptive of gases at room temperature. The use of alumina crucibles may be necessary, however, for samples that can be catalyzed by Pt and for samples that react with Pt to lower its melting point or to swell or embrittle it. DTA and TGA temperature standards certified by the International Confederation for Thermal Analysis are used in thermocouple calibration to 940°C. Drop weights using pure metal wire are used for higher temperatures.

The Inficon IQ200 spectrometer utilizes a Pt-coated single-piece, ceramic mass filter and an electron multiplier. The instrument is capable of monitoring ion peak heights of 10 different masses and recording them on a strip chart. Rapid 'analog' (200-channel) scans for given mass ranges from 1 to 200 amu are also possible. The instrument operates with either constant sensitivity or constant resolution over the entire mass range.

The cold trap and diffusion pump located just beyond the sensor ensure that large quantities of evolved gases can be efficiently removed following gas analysis (Figure 2). The sensor and furnace chamber are joined by a 50-cm curved stainless steel tube. Unfortunately, this relatively long path and the large volume required to enclose the balance within the thermoanalyzer present large surface areas for the adsorption of gases. Except for one short section, the connecting tube can be baked to help remove adsorbed gases. In contrast, instruments designed to optimize EGA by mass spectrometry are free of these limitations and are capable of providing quantitative data by analyzing a direct molecular stream from the sample (Matson *et al.*, 1986). In spite of design and instrumental limitations, the quadrupole spectrometer used at the ROM provides adequate data to interpret TGA curves for most of the minerals we examine.

METHODOLOGY

Sample preparation

The investigation of various sample preparation procedures and their effects on thermal data were summarized by Mackenzie and Mitchell (1970), Smothers and Chiang (1966), and Neumann (1977). The major considerations of most methods are to remove impurities, minimize crystal structure damage due to grinding, control grain size and grain-size distribution, stabilize the sample at a specific humidity prior to analysis, select ideal sample size, maintain reproducible sample packing, match reference to sample, and evaluate the desirability of sample dilution. The first five considerations are relevant to vacuum TGA. TGA techniques at atmospheric pressures are dealt with by Bish and Duffy (1990) elsewhere in this volume.

A sample of the highest possible purity not only provides the most accurate and interpretable data, but is absolutely essential for the analysis of minor and trace volatile constituents. The effectiveness of purification procedures is governed by the initial sample purity, the amount of material available, and the physical properties of the various constituents. Standard purification methods can be employed for large samples, but the separation and manipulation of small samples require special techniques.

Modern microanalytical tools (e.g., the electron microprobe) have put new demands on older techniques such as thermal analysis. In the ROM laboratory, techniques have been developed to prepare smaller and purer samples for TGA-EGA, which will complement microprobe analyses. For coarse-grained samples, manual selection of pure mineral grains may be carried out under the optical microscope. A transparent or translucent vessel allows simultaneous transmitted and reflected illumination of the grains, making both the inclusions and coatings visible in one viewing. Slight crushing or grinding of the grains may be necessary. The sample is normally immersed in methanol (high-performance liquid chromatographic or similar low-residue grade) during the process. If the sample is only slightly impure, the impurities may be segregated and removed with a syringe.

Samples that cannot be satisfactorily purified by handpicking can be centrifuged in heavy-liquid density columns. The method has been successfully applied to samples of >1 mg to <50 mg. The effectiveness of this method decreases for grain sizes <100 μm. The sample is crushed to reduce the number of compound grains and then placed in a clear glass centrifuge tube to which a heavy liquid of appropriate density is added. The tube sides are rinsed with an appropriate solvent. The sample is dispersed with an ultrasonic bath and centrifuged. The layer containing the desired mineral is stirred to decrease the density gradient in this region, and the sample is centrifuged again. Stirring and centrifuging is repeated until the mineral impurities and compound grains are sufficiently separated from the layer of the pure mineral to allow extraction. A positive-displacement pipet tip (0.5 to 1 ml volume) is used manually, i.e., without the pipet, to remove layers of interest. Methylene iodide/acetone is most frequently used for this separation method, but tetrabromoethane/acetone or Clerici solution can also be used. Sodium polytungstate (manufactured by Sometu-Metawo, Berlin) has been proposed as a non-toxic substitute.

The separation of coarse-grained contaminants from suspensions of finer-grained materials is not restricted to clay samples. The only effective method of separating fine-grained brushite, $CaHPO_4 \cdot 2H_2O$, from coarser-grained contaminants of similar density is simply to disaggregate and suspend the sample with ultrasound, and then to recover the suspended brushite from the settled contaminants by decanting or pipeting.

The use of a magnetic separator is limited to the purification of large quantities (>50 mg) of coarse-grained materials, because the loss of material in smaller samples may be unacceptably high. A slightly magnetized needle may be more helpful in handpicking smaller samples.

Once purified, if sufficient quantity of material exists the sample may be more finely ground and the finest suspended fraction decanted and discarded. Sizing a sample in this way is a compromise. A coarse-grained sample has a low surface area and thus has minimal gas adsorption and structural damage produced by grinding, but the diffusion of volatiles is slower and the risk of violent bursting of fluid inclusions is increased. The use of a very fine grained sample speeds diffusion of volatiles from the solid, thereby optimizing TGA results, but adsorption and the proportion of structurally damaged material are increased. Hard samples may be significantly contaminated by grinding; Aines and Rossman (1984) found that the water contents of garnets increased by as much as 0.04 wt. % during grinding. Thus, if possible, a coarser-grained sample is preferred.

If an impurity remains in the purified sample, it should be identified so that the TGA-EGA results can be properly interpreted. A highly exposed film of the X-ray powder diffraction (XRD) pattern of the bulk sample may record enough of the diffraction pattern of the contaminating phase for identification, but if only a trace amount of the contaminant is present, identification may be possible from a Gandolfi or Debye-Scherrer film obtained from a few of the more contaminated grains.

Sample storage

Samples are normally stored at room temperature in a controlled humidity for at least 24 hr prior to analysis. This procedure is essential for a substance suspected of being humidity-sensitive at ambient temperature. Although a humidity level close to the ambient laboratory humidity would minimize changes in water content during weighing, storage in equilibrium with a controlled high humidity provides a closer estimation of water content at saturation.

Saturated aqueous solutions of a variety of salts are useful in providing constant humidity atmospheres (Constant Humidity Table, p. E-46 in Weast, 1980). For suspected humidity-sensitive samples, ammonium chloride is used to provide a 79% level at 23°C. Higher humidities should be avoided as they promote the growth of organic matter in the chamber.

Sample weighing

For most minerals, an approximate volatile content can be estimated on the basis of a low summation of other analytical results or calculated on the basis of an established mineral formula. A sample size can then be chosen so that the rate of weight loss will not likely overload the vacuum system.

A small sample is normally transferred into a vacuum-baked, pre-weighed crucible and weighed so that sample wt. = total wt. - crucible wt. This procedure is necessary, because the complete transfer of powdered samples into a crucible after weighing is impossible. The weighing process must proceed as rapidly as possible, particularly for samples that have been stored at high humidity. The changes in weight due to the loss of adsorbed H_2O from humidity-sensitive material during weighing are monitored for a few minutes, and an initial weight is estimated by extrapolation. Weighing a small sample in a crucible requires a balance having microgram precision or better, large capacity, and sufficient clearance above the pan to accommodate tall crucibles.

At the completion of an analytical run, the sample is cooled in place in the furnace of the thermoanalyzer and brought to atmospheric pressure by filling the vacuum system with dry N_2. When the crucible has cooled to room temperature, it is removed from the thermoanalyzer and immediately weighed. To duplicate the original weighing conditions, the crucible is opened to ambient air for several seconds after the initial weighing and then reweighed. This procedure may increase the weight by a few micrograms for crucible volumes on the order of 1 cm^3. The sample is then monitored for possible further weight gain. If it is found to be adsorbent, the first weight is used in the calculation of total weight loss. The second weight is used if adsorption is negligible.

Vacuum thermogravimetric analysis

The furnace chamber, the spectrometer sensor, the vacuum connection between them, and the crucible (Figure 2) are baked in high vacuum just before the analysis to minimize contamination, especially any 'memory' of volatiles from the previous analysis. The sample is weighed and immediately inserted into the furnace chamber. Contaminating air is then purged from the vacuum system with high-purity N_2 (<2 ppm each of O_2 and H_2O). The system is pumped until the vacuum is better than 1 x 10^{-7} torr. This takes 1-4 hr, depending upon the quantity of loosely held volatiles evolving from the sample. The difference between zero and the position of the beginning of the TGA curve on the ordinate of the thermograms (Figures 1, 3, 5-7, and 9) represents the loss incurred during the initial pumpdown to vacuum at room temperature. Once a steady vacuum is produced, the sample is heated at 10°C/min, but the heating rate may be decreased or increased if the quantity of volatiles is anticipated to be very large or very small. The risks of overwhelming the high vacuum system with an overly rapid volatile loss and of sample ejection caused by rapid volatile loss must be kept in mind. The analysis is normally allowed to continue until no further volatiles evolve or until a major weight loss occurs with no evident increase in any partial pressure, which would indicate that the evolved volatile is condensing on the chamber walls and not reaching the sensor of the spectrometer.

At the end of an analytical run, the sample is cooled, the system is repressurized with N_2, and the sample is quickly reweighed. The total loss determined by TGA is subtracted from the total loss determined by microbalance. The difference represents the weight of loosely bound volatiles lost at room temperature in vacuum prior to analysis. Inasmuch as this value is calculated using three weight measurements, the cumulative error is a relevant factor for samples weighing <5 mg. The method is satisfactory only if the thermal product does not absorb appreciable N_2 or H_2O upon repressurization. For zeolites, heating the sample to temperatures >800°C to destroy the framework structure is advisable.

Because evolving gases are uninhibited by atmospheric pressure, weight losses are accelerated in vacuum. For example, a CO_2 loss by the rare-earth silicate tundrite, $Na_3(Ce,La)_4(Ti,Nb)_2(SiO_4)_2(CO_3)_3O_4(OH)\cdot 2H_2O$ (Karup Moller, 1982), is quite rapid, even at atmospheric pressure; in vacuum this mineral loses 15 wt. % over a 25°C interval at a heating rate of 10°C/min! To avoid sample ejection during such losses, even small samples are normally run in tall crucibles with cocked lids. Crucible walls and lids are checked at the end of a run to determine whether any of the sample may have been ejected. Although any loss of sample would ruin a weight-loss measurement, the temperatures of the weight-loss events and curve shape may still be useful.

If a thermal product is to be identified, the sample can be annealed at the critical temperature for several minutes at the end of the heating run, cooled in vacuum, weighed, and immediately analyzed by XRD. Cooling rates in vacuum are relatively slow; hence, samples may quench poorly. Glass capillaries can be used to encase humidity-sensitive products for analysis by XRD.

Evolved gas analysis

The quadrupole mass spectrometer used in the ROM system can be operated in two modes: continuous scanning over any range between mass 1 and 200 and continuous monitoring of as many as 10 pre-programmed channels. All components of the instrument that can be baked are maintained at 100°C during analysis to minimize adsorption of evolving gases on the exposed surfaces. In mineral analysis six masses, namely, 2 (H_2^+), 18 (H_2O^+), 28 (N_2^+, CO^+), 32 (O_2^+), 44 (CO_2^+), and 64 (SO_2^+), are routinely monitored along with any other anticipated masses. Analysis proceeds with periodic scans over the full mass range, particularly near the onset and the peak of a weight loss, to determine if any new volatile species are evolving. A characteristic mass for each new volatile is selected, entered on a channel, and continuously monitored. On the present ROM equipment this procedure is done manually, which causes a break in data collection of 1-2 min while a scan is collected and evaluated. Under software control, the spectrometer may resume routine monitoring immediately following the full spectrum scan, while the software performs the full spectrum analysis and selects the additional masses to be monitored.

The peak corresponding to the parent molecule is usually monitored, unless a characteristic fragment is much more abundant. For example, SO_3 produces a very weak peak relative to the SO_2^+ fragment due to the high degree of dissociation under these analytical conditions (Langer and Gohlke, 1967). Here, the SO_2^+ fragment is monitored.

A further consideration is that a peak for a given mass may not be unique to a specific volatile. For example, mass 32 may be O_2^+, $^{32}S^+$, or a mixture of the two. To resolve this, the mass 34/32 ratio must be determined. This ratio is 0.002 for O_2 and 0.042 for S. If the ratio indicates the presence of S, the mass spectrum is examined for the characteristic peaks of H_2S, SO_3 (inferred by the presence of SO_2), or S_8. Incidentally, a sulfur vapor spectrum, such as given by Langer and Gohlke (1967), might be misinterpreted as an SO_2 spectrum should it be recorded by a mass spectrometer having a low mass range (1-50 amu).

Instruments having components that cannot be baked (e.g., the balance chamber in the ROM instrument) will likely have H_2O as the most abundant residual gas, followed by N_2, O_2, CO_2, and H_2. As a consequence of the high H_2O background, relatively large samples (10-50 mg) must be analyzed to detect trace or minor water (0.X - 0.0X wt. %). Furthermore, the cracking pattern for water shows abundant fragments at masses 16 (O^+) and 17 (OH^+). These masses interfere with the peaks for methane and ammonia parent molecules, respectively, so the less-abundant fragments CH^+ and NH^+ must be monitored if these gases are present. The presence of residual hydrocarbons from vacuum pumps and seals typically contribute a minor peak at mass 40 which raises the background for Ar. Robinson (1968) discussed the interaction of residual gases with contaminants of ionization filaments (ionization gauges and spectrometer ion sources) to produce elevated levels of CO and other gases.

Mass spectra are identified by comparing experimental results with databases provided by the manufacturer or with the graphical compilations of McLafferty and Stauffer (1989), or with numerical data of Cornu and Massot (1975). The listed inorganic spectra are incomplete; hence, the user may find it necessary to obtain spectra from reference gases or other substances.

Evolved gas curves presented in this chapter (Figures 1, 3, 5-7, and 9) represent ion peak heights expressed in arbitrary units for either parent molecules or fragments thereof. Where multiple curves are presented at different scales, the relative amplitude factors are given for each. Note that peak heights reflect but do not necessarily represent relative molecular abundances.

INTERPRETATION

The major factors to be considered during the interpretation of vacuum TGA-EGA results are illustrated below using results from studies in the ROM laboratory and from the literature. In the routine analysis of evolved volatiles, the partial pressure curves of each gas must be correlated with the TGA curve so that weight losses can be assigned to each volatile. This procedure is straightforward for an analysis involving a single volatile, because the total pressure curve matches the evolved gas curve. Figure 3 shows the TGA and total pressure curves for vermiculite, $(Mg_{2.54}Fe^{2+}_{0.08}Fe^{3+}_{0.40}Ti_{0.50})(Si_{2.76}Al_{1.24})O_{10}(OH)_2[Ca_{0.02}Mg_{0.28}]^{0.60+}$, from Kenya. This material has been extensively studied, most recently by Brydon and Turner (1972) who found it to be free from both mica and chlorite. We examined the <2-μm size fraction of the Brydon and Turner sample, stabilized at 40% RH and 20°C prior to analysis. The sample was placed in the thermoanalyzer and pumped for 100 hr to a vacuum of 1.5 x 10^{-7} torr, during which it lost 9.4 wt. %. The lower temperature (<300°C) regions of the TGA and total pressure curves display three interlayer water-loss peaks at 80°, 100°, and 185°C, which are characteristic of vermiculite. The peak profile of the total pressure curve is similar to the profile of the DTA curve of Walker and Cole (1957) for Westchester vermiculite equilibrated at 0% RH. The 80°C peak on both the DTA curve of Walker and Cole and on our total pressure curve is weak, indicating that the recorded 9.4 wt. % loss during the pumpdown prior to heating was the loss of weakly bonded interlayer water.

In the mid-temperature region (300°-600°C), Brydon and Turner (1972) recorded a single endotherm, having a peak at 508°C, and a very weak endotherm at 585°C (Figure 4), on their vacuum DTA curve. In contrast to this result, the main peak on the total pressure curve (Figure 3) is at 420°C followed by weak peaks at

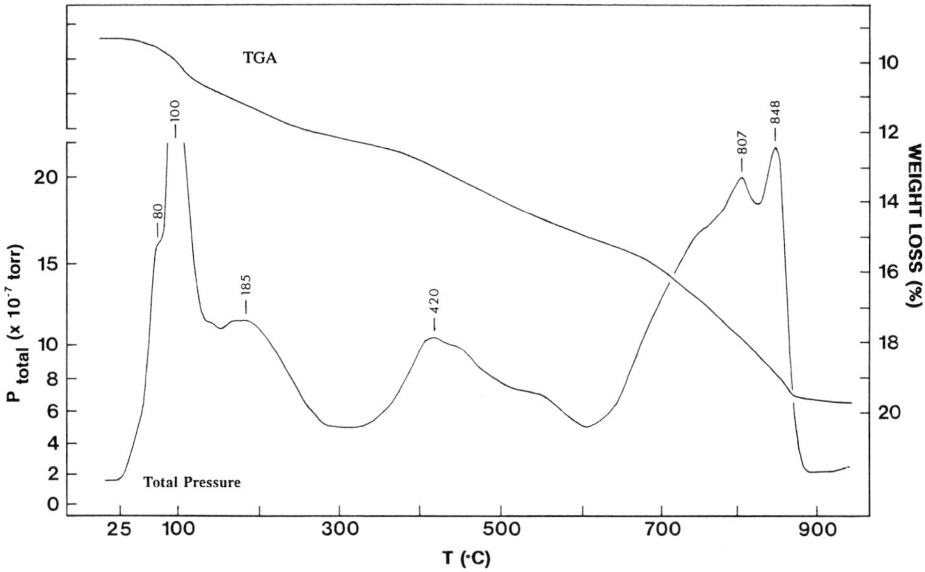

Figure 3. Thermal gravimetric analysis and total pressure curves of vermiculite, Kipiponi, Kenya, (ROM sample #M32300), (10.4 mg), heated at 10°C/min, initial vacuum 1.5 x 10^{-7} torr.

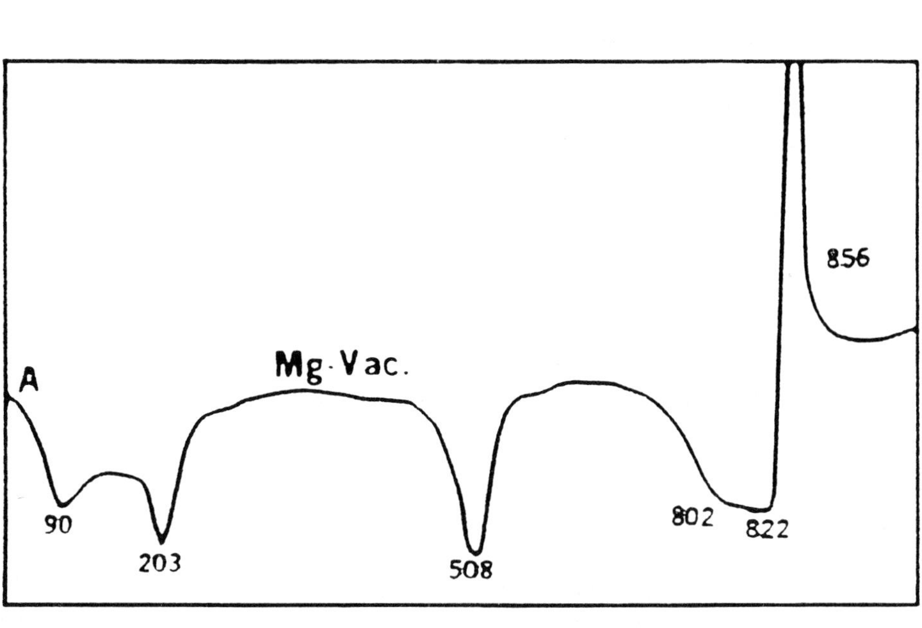

Figure 4. Differential thermal analysis curve of vermiculite, Kipiponi, Kenya, <5-μm size sample, heated at 10°C/min, vacuum 0.5 mm Hg (0.5 torr). From Brydon and Turner (1972).

455° and 555°C. Brydon and Turner (1972) used a modest vacuum of 0.5 mm Hg (0.5 torr) for their DTA run, so a direct comparison of the two sets of results is difficult. The multiple midrange peaks in Figure 3 either could indicate the presence of other phyllosilicates, but because Brydon and Turner found none, the total pressure curve probably instead illustrates the complex, multi-stage nature of the dehydroxylation reaction. An increased sensitivity to fine details, such as these, is typical of high-vacuum TGA-EGA studies.

In the high-temperature region (600°-900°C), the shape of the total pressure curve indicates that dehydroxylation continues above 600°C, and water-loss peaks can be seen at 807° and 848°C in Figure 3. The 807°C peak corresponds with the 822°C endotherm recorded by Brydon and Turner (Figure 4), but it is in part produced by a minor loss of CO_2 (not shown), possibly from a minor impurity. The 848°C peak is related to the final loss of hydroxyl during recrystallization and corresponds well with the 856°C exotherm recorded by Brydon and Turner.

Should the division of a complex weight loss from a single volatile species into steps be necessary, points of minimal partial pressure may be used in the same manner as, but more precisely than, DTG minima. The division of the losses at a total pressure minimum or a DTG minimum would probably result in an erroneously high value for the first volatile and a correspondingly low value for the second because weight losses typically commence more gradually than they terminate. Thus, the rapidly decreasing loss due to one volatile compound may overlap the gradually increasing loss of a second. Because such asymmetry is evident in most evolved gas curves, a more accurate temperature can be estimated from the point at which the tail area of the first volatile is equivalent to the onset area of the second volatile, with consideration being given to relative ionization probabilities, yields, and masses.

If losses of two volatiles are strongly overlapping or simultaneous, only the combined weight-loss percentage normally can be expressed, although a ratio may be approximated from peak areas on evolved gas curves. For minor partial pressure peaks, the proportion of that volatile can be estimated to within an order of magnitude. If very minor peaks are of interest, their source needs to be determined because they may: (1) originate from the sample, (2) be an artifact of adsorbed gases being stripped from system surfaces by molecules of the principal volatile, or (3) have formed by reaction among evolved gases, residual gases, and system surfaces. Comparative analyses of appropriate characterized materials will likely be necessary to solve this type of problem.

The majority of hydrated minerals evolve H solely as H_2O (Figure 3); however, a significant number of minerals have been found to evolve H_2 and HF in addition to H_2O, but rarely hydrocarbons or HCl. The presence of other H-bearing species may seriously complicate an otherwise straightforward analysis for water. The H_2 loss at 360°C from maricopaite, $Pb_{7.00}Ca_{2.17}(Si_{34.46}Al_{11.34})O_{97.10}\cdot 30.96H_2O$, (Peacor *et al.*, 1988) was unrelated to the H_2O loss at 100°C or the water loss at

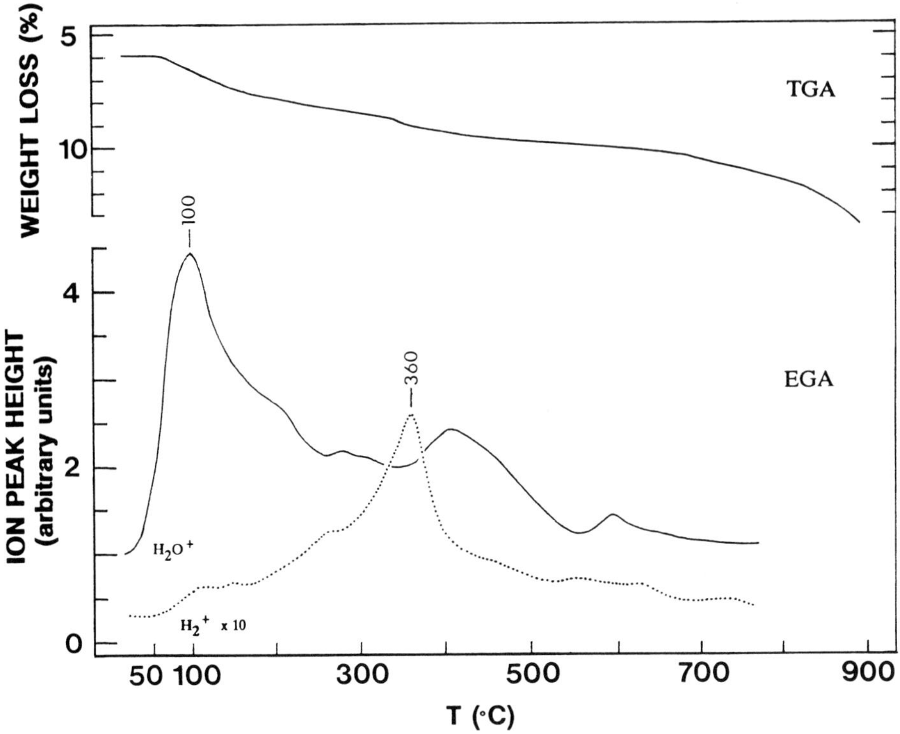

Figure 5. Thermal gravimetric-evolved gas analysis curves of maricopaite, Tonopah, Arizona, (2.54 mg), heated at 10°C/min, initial vacuum 2 x 10^{-8} torr.

410°C (Figure 5). The source of the H_2 was unknown. This problem was complicated in that most of the available material (2.54 mg) was used for the first thermogram. Fortunately, an additional small amount of sample was made available; a pure 0.565-mg sample was obtained by density gradient separation in a centrifuge. The repeat analysis indicated that the source of the H_2 was hydrocarbons presumably occluded within the zeolite-like cages in the structure.

Hydrogen is commonly lost from hydrous Fe-bearing minerals. The theromogram (Figure 6) of the unusual ferroan talc, described by Robinson and Chamberlain (1984), $(Mg_{1.90}Fe^{2+}_{0.78}Fe^{3+}_{0.26}Al_{0.02}\,\square_{0.04})Si_{4.04}O_{10.06}(OH)_{2.23}$, illustrates this phenomenon. Hydrogen begins to evolve at 500°C and peaks at 680°C. Water produced by the dehydroxylation of the talc begins to evolve at 600°C and peaks at 755°C. The second small H_2^+ peak at 755°C on the main H_2^+ peak is a cracking fragment of the H_2O which produces the H_2O^+ peak of 755°C. The evolution of H_2 appears related to a local auto-oxidation-dehydroxylation reaction involving Fe^{2+} and OH. Dehydroxylation is followed by an O_2 loss over 1000°C due to the re-duction of Fe^{3+} (shown only on the TGA curve on Figure 6). The dehydroxylation peak temperature in talc is inversely proportional to iron content: 755°C for the ferroan talc (16.8 wt. % total Fe as FeO, Figure 6), 845°C for moderate-iron talc (2.3 wt. %

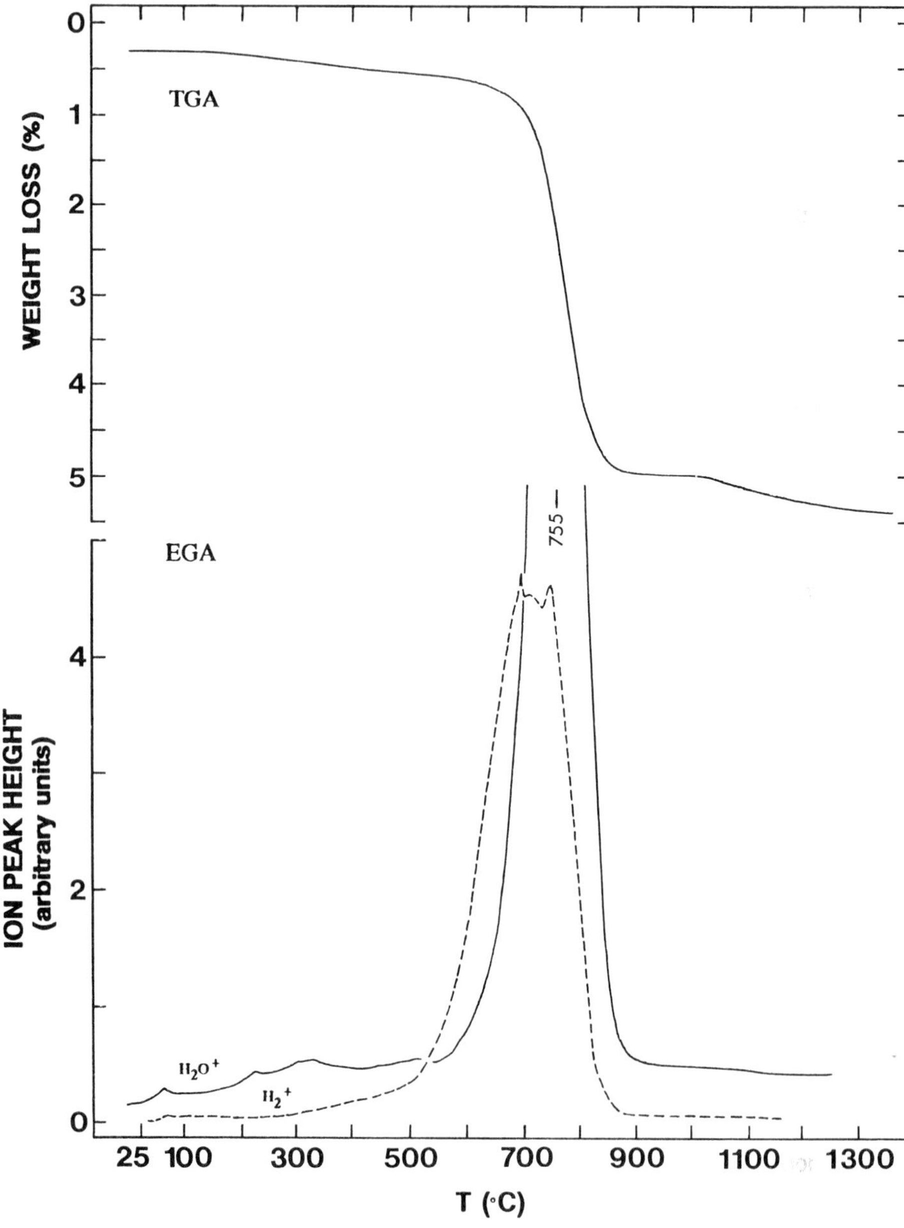

Figure 6. Thermal gravimetric-evolved gas analysis curves of ferroan talc, Sterling mine, Antwerp, New York, (15.5 mg), heated at 10°C/min, initial vacuum 1 x 10^{-7} torr.

total Fe as FeO) from Frontenac County, Ontario (ROM #362), and 884°C for a low-iron talc (<1 wt. % FeO) from North Carolina (ROM #M6459). Evolved H_2 was not detected in TGA-EGA of the latter two talc samples.

A distinct H_2 loss, accompanied by F_2, HF, and H_2O losses, is also a feature of F-bearing hydrosilicates, such as apophyllite, astrophyllite, biotite, and amphibole, and even relatively anhydrous minerals, such as titanite. Astrophyllite, biotite, and amphiboles show similar overlapping losses of H_2, H_2O, and HF with partial pressure peaks appearing in that order, or for some samples, with H_2 and H_2O peaks superimposed. The thermogram of a biotite (lepidomelane) (Figure 7) illustrates the complexity of this type of thermal decomposition. Between 500° and 900°C the H_2 loss is independent of the H_2O loss in the same region, but coincides with a smaller, detectible F_2 loss. HF does not evolve in the 500°-900°C interval, but evolves at temperatures >850°C. At high temperature (>1100°C), fluorides, such as SiF_4, are lost.

Studies on F_2 losses from a series on titanite samples (Hawthorne *et al.*, 1990) indicate that F in titanite is a major factor controlling the nature of the gases that evolve between 500° and 1000°C. If F is present in titanite, H_2 is the principal evolved gas; whereas if F is not present, H_2O is the principal evolved gas. Although many, but not all, of the titanite samples exhibiting this feature are Fe-bearing, Fe is not a key element in this decomposition. At higher temperatures (1000°-1325°C), SiF_4, F_2, HF, H_2, and O_2 are lost. Figure 8 indicates that the weight loss between 1000°C and 1325°C is directly proportional to the amount of F present in titanite. The reactivity of F at high temperatures between evolved F_2 and residual H_2 or H_2O

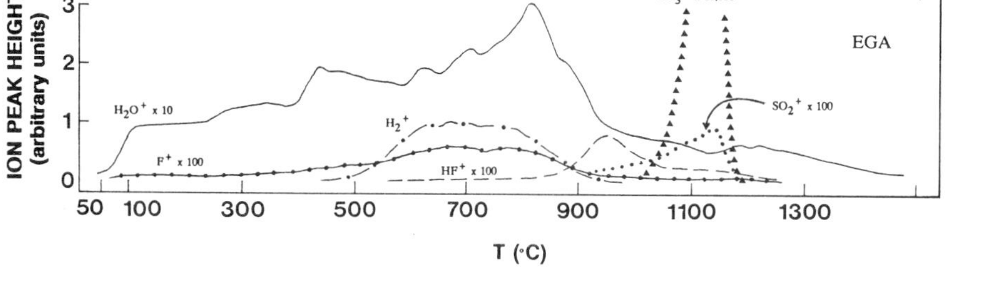

Figure 7. Thermal gravimetric-evolved gas analysis curves of F-bearing biotite, Faraday Township, Ontario (ROM #M14701) (34.0 mg), heated at 10°C/min, initial vacuum = 1 x 10^{-7} torr.

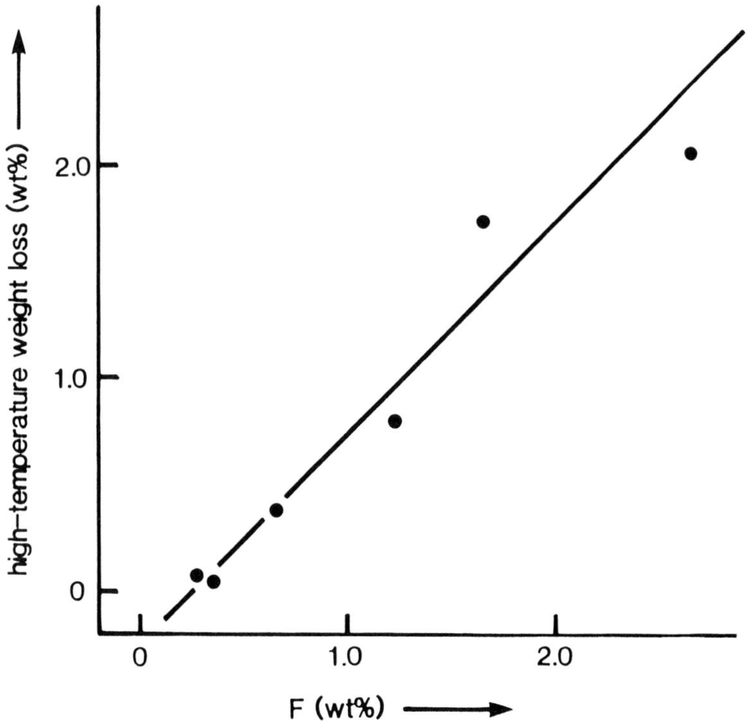

Figure 8. Plot of high temperature (>1100°C) weight loss (wt. %) vs. F content (wt. %) of titanite specimens from various localities. From Hawthorne *et al.* (1990).

was discussed by Nikolaev *et al.* (1972). Further work is required to understand the mechanism of F loss.

The evolution of HCl from hilgardite, $Ca_2B_5O_9Cl{\cdot}H_2O$, (Rachlin *et al.*, 1986), between 950°and 1200°C after the loss of H_2O between 420°and 640°C raises the question of the source of the H that forms the HCl. Some of the HCl possibly formed in a high-temperature reaction between residual H_2 or H_2O and evolved Cl, similar to the reaction of F discussed by Nikolaev *et al.* (1972). Because some F-bearing and Cl-bearing volatiles may be condensable, a sample cannot be considered F- or Cl-free simply because no volatiles were observed by EGA.

Many minerals contain very weakly bonded volatile components, which may be lost by exposure to high vacuum at room temperature. The mineral weddellite, $Ca(C_2O_4){\cdot}2.26H_2O$, is an extreme case. It loses 91 wt. % of its H_2O under such conditions (Mandarino and Witt, 1983). With the ROM instrument, these room-temperature losses could only be determined indirectly as the difference between the TGA loss recorded during heating and the total loss determined by microbalance at the end of the run (i.e., sample wt. - product wt.). If a significant loss takes place at room temperature, a further loss invariably occurs as heating commences, which can be studied by EGA. Presumably these volatiles emitted during initial heating are representative of those lost during pump-down. To date, H_2O has been found to be the predominant loosely held volatile, although other minor components may be present. Small quantities of acetone were identified in low-temperature losses from kraisslite, $(Mn^{2+},Mg)_{24}Zn_3Fe^{3+}(As^{3+}O_3)_2(As^{5+}O_4)_3(SiO_4)_6(OH)_{18}$, but the acetone was most likely an artifact of preparation.

The decomposition of arsenates and arsenites is marked by the detection of $As_2O_3^+$ (and numerous other fragments), as well as $AsCl_2^+$ from Cl-bearing phases. The breakdown of Cl-bearing borates produces BCl_3^+ and $CaCl_4^+$ and their fragments and, possibly, BO^+.

The study of oxalates may be complicated by the formation of elemental carbon during thermoanalysis. If the thermal product of calcium oxalate is a medium gray, due to residual carbon instead of the expected white of CaO, the analysis may be completed by heating the product in O_2 to oxidize the C.

The interpretation of the thermoanalysis of sulfides is assisted by the large volume of literature available on the phase relations of sulfides; however, the phase diagrams represent equilibrium conditions and the thermograms represent results of a dynamic, non-equilibrium system, with the volatiles constantly being pumped from the sample as temperature increases. Thus, the bulk composition may be in a state of continuous change. The interpretation of the TGA-EGA results of millerite, NiS, (Figure 9), using the phase diagrams of Kullerud and Yund (1962) for the system Ni-S in the presence of vapor (Figure 10), is a typical example. Although S is first lost in small amounts at 410°C (not shown), the major losses occur in two steps, with S pressure maxima at 713°C and 1142°C (Figure 9). A comparison of the TGA-EGA data with the phase diagram suggests a path from NiS to $Ni_{3-x}S_2$ + αNiS (field 11 in Figure 10) and $Ni_{3\pm x}S_2$ through the first 713°C weight loss. A liquid component appears just below 800°C (field 4, Figure 10), and the gradual loss from 800° to 1000°C is produced by S being pumped from the vacuum system, which keeps the S vapor pressure of the Ni-S system in disequilibrium. At 880°C the S vapor pressure is at a minimum and the bulk composition is Ni_5S_3. The second major S loss at 1142°C occurs from a melt (field 2 = L + Ni_{ss}, Figure 10). Sulfur was detected as S^+_n, where n = 1 to 6, all apparently being ionic fragments of the S_8 molecule. The persistence of a significant S vapor pressure >1200°C after the second major weight loss suggests that some of the S condensed on the walls of the vacuum system en route to the sensor and was then being pumped off as the S vapor pressure dropped.

Volatiles having very low vapor pressures may condense in the coolest zone en route to the spectrometer and will not be detected. Such condensation has been noted particularly for Na-, Pb-, and B-bearing minerals. Graziani and Lucchesi (1982) described the loss of significant amounts of Na and K from scapolites in the form of NaCl and KCl. To prevent condensation of these chlorides, the vacuum passageway must be heated to several hundred degrees.

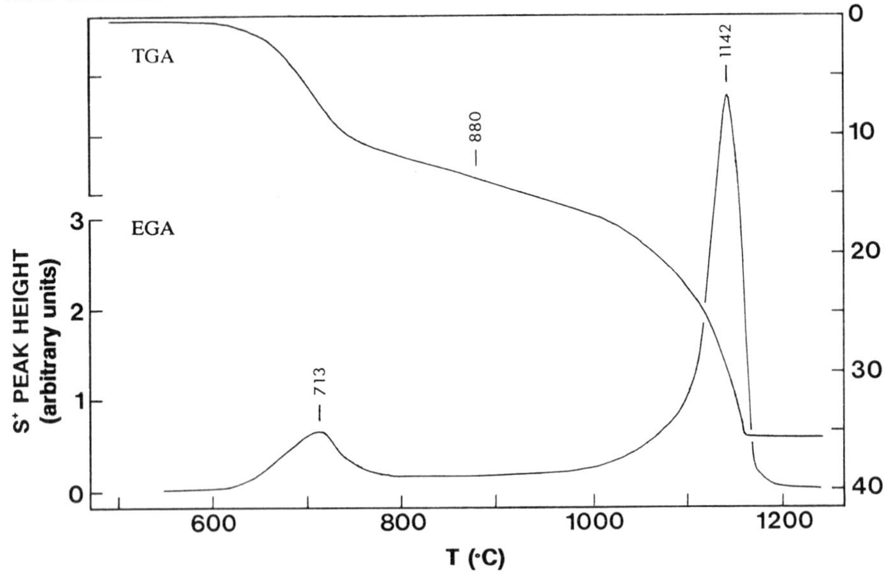

Figure 9. Thermal gravimetric-evolved gas analysis curves of millerite, Gap mine, Pennsylvania (ROM #E1888) (11.7 mg) heated at 10°C/min., initial vacuum 1 x 10^{-7} torr.

Although condensing volatiles are generally restricted to high temperatures following losses of volatiles, such as H_2O and CO_2, the same problem has also been encountered at lower temperatures. In one experiment, a phosphate of composition $(Mn,Fe^{3+},Fe^{2+},Na,Ca)_7Ca_2(PO_4)_6(OH,H_2O)_4$, which is likely the Mn-analog of wicksite (Peacor *et al.*, 1985), underwent a rapid weight loss during which H_2O was the only major volatile detected. The total loss was considerably greater than anticipated. Upon reexamination the rate of loss was found to have increased dramatically for a narrow 50°C interval within an otherwise normal weight loss. This rapid loss, however, was not reflected in the partial pressure curves of water or of any other detectable volatile. Apparently, this loss was produced by a co-evolving volatile that condensed on the walls of the vacuum system prior to reaching the quadrupole mass spectrometer; hence, it could not be identified.

The possibility of interaction of components during the analysis of mixtures must be considered. In a study of brucite-calcite mixtures as possible standards in EGA, we observed that the water loss for brucite was not quite completed at the expected temperature and the remaining water appeared to be lost during the higher-temperature decarbonation of calcite. The extent of this water 'retention' was roughly proportional to the sample size, suggesting that intermediate compounds may still form from mixtures in vacuum, as they do at atmospheric pressure. Such a reaction at atmospheric pressure between water vapor and calcite prior to decomposition was discussed by Langer and Gohlke (1967, their Figure 3.14).

If DTA is conducted in vacuum, events involving transitions appear somewhat weaker than if they are measured at atmospheric pressure due, in part, to lower

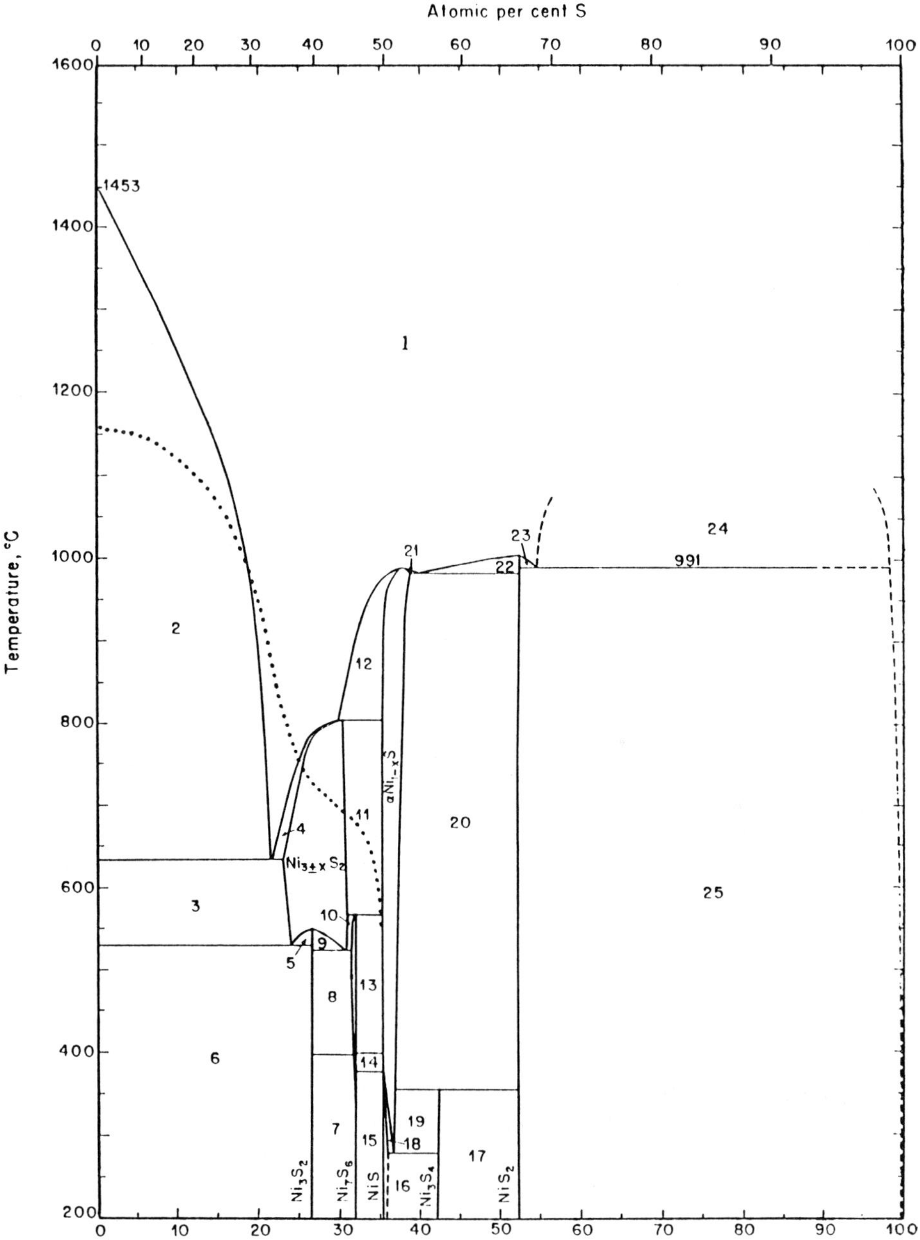

Figure 10. "Condensed" phase diagram of system Ni-S >200°C, from Kullerud and Yund (1962). Field 11 = $Ni_{3-x}S_2$ + αNiS; field 4 = L + $Ni_{3\pm x}S_2$; field 2 = L + Ni_{SS}. Thermogram of Figure 9 is plotted as a dotted line.

thermal conductivity in vacuum. The endotherms involving loss of volatiles are further weakened by the efficient removal of the evolving gas by the vacuum pumps.

SUMMARY

Advantages of the vacuum method

1. If a thermoanalyzer is operated in vacuum, the TGA baseline is flat and the `buoyancy' effect present during analysis at atmospheric pressure is eliminated. The buoyancy effect is caused by the convection and the reduction in density of the gas around the crucible as it is heated, which produces an apparent weight gain. Although compensation for the buoyancy effect is possible by subtraction of a TGA curve for an equivalent volume of inert material from the TGA curve of the sample analyzed, the error may be large at lower temperatures where the effect is most pronounced. Most modern microbalance designs have reduced or eliminated this problem.

2. At pressures $<10^{-4}$ torr, diffusion from sample surfaces is effectively uninhibited, and volatiles are efficiently removed from the sample during decomposition. This efficient removal of volatiles reduces the influence of sample size on decomposition temperatures and improves the reproducibility and comparability of successive analyses.

3. Evolved gases are easily and efficiently directed from the furnace to the gas analyzer without dilution so as to produce a relatively high sensitivity and instantaneous response for EGA. The problems introduced by impurities in a carrier gas are eliminated.

4. The oxidation of a sample can be controlled in vacuum. Ferrous compounds tend to oxidize even at moderate vacuum (10^{-3} torr), causing gradual weight gains at moderate temperatures (400°-600°C). The O_2 causing this may originate from air leaks in the system, incomplete flushing of air from the sample, or minor O_2 impurities in a flushing gas. At pressures $<10^{-6}$ torr, however, oxidation does not occur in minerals that are otherwise prone to oxidation. (An atmosphere containing 5% H_2 would have a similar effect, but analysis for H would not be possible.)

5. The use of conventional H_2O- and H_2O+ terms to express the state of water in an analysis may serve a purpose in rock analysis, but it may well be misleading or uninterpretable when applied to mineral analysis. An expression of water content in terms of H_2O lost and H_2O retained in vacuum is generally more interpretable and is particularly of value in interpreting electron microprobe analyses of hydrated minerals. It would be easy to design a simple vacuum system for determining H_2O lost and H_2O retained in vacuum for whole rock analysis.

6. Sorption-desorption and reactivity of vacuum-heated solids can be conveniently studied. A sample can be heated in vacuum to a specific temperature, a gas can be admitted into the vacuum system for a given period, the sample can be cooled, and TGA-EGA can then be run, all on the same instrument.

7. Operation in vacuum reduces the importance of grain size and packing so that comparative analyses are much easier to make. For example, calcite thermograms in vacuum typically shows decomposition peaks in a narrow range of 650°-665°C, whereas in atmosphere the range is considerably greater (860°-1010°C; Webb and Heysteck, 1957) due to the effect of differences in grain size and packing.

8. The recorded total pressure and partial pressure curves of evolved gases in vacuum analysis are less noisy and, hence, more sensitive than similar curves recorded from analysis in atmosphere or from other techniques, such as DTG.

9. Detection of H_2, F_2, and other reactive volatiles may be possible only in vacuum studies. These volatiles generally react with other gases or the walls of the instrument during analysis at atmospheric pressure prior to transfer to the gas analyzer.

Disadvantages of the vacuum method

1. The weight loss of volatiles from a sample as the vacuum system is being pumped down cannot be directly measured, and the species causing the loss cannot be identified with the mass spectrometer. Therefore, the detailed study of intermediate hydrates involving loosely bound volatiles requires analysis at atmospheric pressures.

2. Major fluctuations in evolved gas pressure during decomposition will affect the TGA baseline of some balances in vacuum. For many materials, this may not be a problem because critical measuring points in most weight-loss determinations lie before, between, and after weight-loss events and, thus, are on the baseline and are unaffected by pressure changes. A weight value at the dividing point between poorly resolved losses, however, will be affected. Normally, a correction can be applied to this value based on the pressure of the volatile at that point (see Oswald and Wiedemann, 1977, p. 159). If accurate weight values are required throughout the analysis, as in studies of kinetics, the balance must be of a design unaffected by pressure changes. If this is not possible, the effect might be minimized by the use of a constant leak (He, Ar, or N_2), so that the residual gas pressure would be about an order of magnitude greater than any evolved gas pressures. Alternatively, a slower heating rate might help solve or minimize the problem.

3. Weight losses occur at lower temperatures in vacuum particularly in the higher-temperature range; thus, much more overlap of weight losses may exist than in

studies in atmosphere. For example, water determinations of arsenates and F-rich phosphates and silicates are commonly complicated by a considerable overlap of H_2O, arsenate, and fluoride volatiles. At one atmosphere, the same losses may show less or even negligible overlap.

4. The condensation of some volatiles lost at high temperature before they reach the detector will produce an incomplete analysis. Use of a direct molecular beam will solve this problem, or, for some materials, the volatiles might be identified later if they can be collected on a metal or ceramic plate that has been positioned in a cool section of the furnace.

5. Each vacuum analysis is time-consuming. The bake-out required to remove volatiles from the walls of the vacuum system and the pump-down time necessary to stabilize a high vacuum add considerably to the time of each analytical run. Consequently, this technique is not a rapid, high-volume analytical procedure. Samples are chosen carefully so that the results are likely to be worth the time and effort. A high-temperature analysis (1450°C) typically requires two days of effort, from sample preparation to report, which is considerably longer than a routine thermal analysis. Computer control and automation will reduce interpretation and graphics production time, but not the bake-out and pump-down time. The use of medium-temperature furnaces (1000°C), which cool rapidly, shortens analytical time to about one day.

6. Only a small proportion of the thermoanalytical literature deals with any aspect of thermal analysis of minerals in vacuum. The literature dealing with analyses in atmospheric pressure is frequently incomplete (for example, the atmosphere may not be specified), limiting the use of this material for comparative studies with vacuum results.

CONCLUSIONS

Vacuum TGA-EGA is a useful analytical technique that produces results not easily obtained by other methods. The improvement of quadrupole mass spectrometers to analyze for H has opened an important area of study for hydrated minerals and other materials. The mechanisms of decomposition and evolution of gas, particularly H_2 and F_2, are not well understood, and further work is required on these problems.

Computer control of the operation and data collection of TGA-EGA instrumentation speeds the operation of the equipment, greatly improves the accuracy of the analysis, and increases the capacity for interpretation of the results. Although not a routine analytical tool, TGA-EGA can provide important information in mineral studies.

ACKNOWLEDGMENTS

The authors thank R. F. Giese, Jr., State University of New York at Buffalo, who invited us to present this review at the CMS Workshop held at Socorro, New Mexico in September 1987; D. R. Pevear, Exxon Production Research, Houston; and Stephen Guggenheim, University of Illinois at Chicago, for advice and encouragement with this project. The authors acknowledge with thanks the support of the Natural Sciences and Engineering Research Council of Canada for operating and equipment grants to Fred Wicks, through his cross-appointment to the Department of Geology, University of Toronto, that helped support the operation of the thermoanalyzer at the Royal Ontario Museum.

We acknowledge with thanks *Clays and Clay Minerals*, the *American Mineralogist*, and the *Journal of Petrology* for permission to use Figures 4, 8, and 10 respectively.

REFERENCES CITED

Aines, R. D. and Rossman, G. R. (1984) The hydrous component in garnets: pyralspites: *Amer. Mineral.* **69,** 1116-1126.

Bish, D. L. and Duffy, C. J. (1990) Thermogravimetric analysis of minerals: in *CMS Workshop Lectures, Vol. 3, Thermal Analysis in Clay Science,* J. W. Stucki, D. L. Bish, and F. A. Mumpton, eds., The Clay Minerals Society, Boulder, Colorado, 95-157.

Breth, A., Dobrozemsky, R., and Schwarzinger, G. (1983) Some basic problems in the deconvolution of gas-analytic mass spectra: *Int. J. Mass Spectrometry Ion Physics* **48,** 3-6.

Brydon, J. E. and Turner, R. C. (1972) The nature of Kenya vermiculite and its aluminum hydroxide complexes: *Clays & Clay Minerals* **20,** 1-11.

Clinckemaille, A. and Hofmann, C. (1972) Application of a scan-programmer to control a quadrupole mass spectrometer: in *Thermal Analysis, Proc. 3rd Int. Conf. Thermal Analysis, Davos, Switzerland, 1971, Vol. 1,* H. G. Wiedemann, ed., Birkhauser Verlag, Basel, Switzerland, 337-342.

Corna, A. and Massot, R. (1975) *Compilation of Mass Spectral Data:* 2nd ed., Heyden, London, 2 volumes.

Dawson, P. H., ed. (1976) *Quadrupole Mass Spectrometry and its Applications:* Elsevier, Amsterdam, 349 pp.

Dollimore, D., Gamlen, G. A., and Taylor, T. J. (1984) Mass spectrometric evolved gas analysis—an overview: *Thermochim. Acta* **75,** 59-69.

Dunner, W. and Eppler, H. (1975) Advanced coupling systems for thermoanalyzer with quadrupole mass spectrometer: in *Thermal Analysis, Proc. 4th Int. Conf. Thermal Analysis, Budapest, 1974, Vol. 3,* I. Buzas, ed., Heyden, London, 1049-1060.

Elliot, R. M. (1969) Single focusing magnetic deflection mass spectrometers: in *Mass Spectrometry, Proc. Symp. Mass Spectrometry, Enfield College of Technology, 1967,* R. Brymner and J. R. Penney, eds., Chemical Publishing Co., New York, 3-14.

Gallagher, P. K. (1982) Applications of evolved gas analysis to the study of inorganic materials and processes: *J. Thermal Anal.* **25,** 7-20.

Graziani, G., and Lucchesi, S. (1982) The thermal behavior of scapolites: *Amer. Mineral.* **67,** 1229-1241.

Green, G. W. (1968) *The Design and Construction of Small Vacuum Systems:* Chapman and Hall, London, 181 pp.

Harris, D. M. (1981) The microdetermination of H_2O, CO_2, and SO_2 in glass using a 1280°C microscope vacuum heating stage, cryopumping, and vapor pressure measurements from 77 to 273 K: *Geochim. Cosmochim. Acta* **45,** 2023-2036.

Hastie, J. W. (1979) Characterization of high temperature vapors and gases: *Proc. 10th Materials Research Symp., Gaithersburg, Maryland, 1978, Vol. 1, Natl. Bur. Stand. Spec. Publ.* **561,** 812 pp.

Hawthorne, F. C., Groat, L. A., Raudsepp, M., Ball, N. A., Kimata, M., Spike, F. D., Lumpkin, G. R., Ewing, R. C., Gregor, R. B., Lytle, F. W., Ercit, T. S., Rossman, G. R., Wicks, F. J., Ramik, R. A., Sherrif, B. L., Fleet, M. E., and McCammon, C. (1990) Alpha-decay damage in natural titanites. *Amer. Mineral.* **75** (in press).

Holdiness, M. R. (1984) Evolved gas analysis by mass spectrometry: A review: *Thermochim. Acta* **75,** 361-399.

Kaisersberger, E. and Emmerich, W. D. (1982) Theoretical and instrumental studies for the coupling of TA systems and quadrupole mass spectrometers: in *Proc. 7th Int. Conf. Thermal Analysis,* B. Miller, ed., Wiley, New York, 279-283.

Karup Moller, S. (1982) Tundrite from the Ilimaussag alkaline intrusion, South Greenland: *N. Jb. Mineral. Mh.*, 481-494.

Kullerud, G. and Yund, R. A. (1962) The Ni-S system and related minerals: *J. Petrology* **3,** 126-175.

Langer, H. G. (1986) Early developments of EGA by mass spectrometry: *Thermochim. Acta* **100,** 187-202.

Langer, H. and Gohlke, R. S. (1967) Mass spectrometric identification of gaseous products from thermal analysis: in *Gas Effluent Analysis,* W. Lodding, ed., Marcel Dekker, New York, 71-100.

Low, M. J. D. (1967) Analysis of gas effluent streams by infrared absorption: in *Gas Effluent Analysis,* W. Lodding, ed., Marcel Dekker, New York, 155-193.

Mackenzie, R. C. and Mitchell, B. D. (1970) Technique: in *Differential Thermal Analysis, Vol. 1,* R. C. Mackenzie, ed., Academic Press, London, 101-122.

Mandarino, J. A. and Witt, N. V. (1983) Weddellite from Biggs, Oregon, U.S.A.: *Can. Mineral.* **21,** 503-508.

Matson, D. W., Muenow, D. W., and Garcia, M. O. (1986) Volatile contents of phlogopite micas from South African kimberlite: *Contrib. Min. Pet.* **93,** 399-408.

McLafferty, F. W., and Stauffer, D. B. (1989) *The Wiley/NBS Registry of Mass Spectral Data,* Wiley, New York, 7 volumes.

Morgan, D. J., Warrington, S. B., and Warne, S. St. J. (1988) Earth science applications of evolved gas analysis: a review: *Thermochim. Acta* **135,** 207-212.

Neumann, B. S. (1977) Thermal techniques: in *Physical Methods in Determinative Mineralogy,* J. Zussman, ed., Academic Press, London, 605-662.

Nikolaev, N. S., Suvorova, S. N., Gurovich, E. I., Peka, I., and Korchemnaya, E. K. (1972) *Analytical Chemistry of Fluorine:* Halsted Press, New York, 222 pp.

Oswald, H. R. and Wiedemann, H. G. (1977) Factors influencing thermoanalytical curves: *J. Thermal Anal.* **12,** 147-168.

Peacor, D. R., Dunn, P. J., Ramik, R. A., Campbell, T. J., and Roberts, W. A. (1985) A wicksite-like mineral from the Bull Moose Mine, South Dakota: *Can. Mineral.* **23,** 247-249.

Peacor, D. R., Dunn, P. J., Simmons, W. B., Wicks, F. J., and Raudsepp, M. (1988) Maricopaite, a new hydrated Ca-Pb silicate from Arizona: *Can. Mineral.* **26,** 309-313.

Prime, R. B. (1987) New techniques, news and applications: *NATAS Notes* **19,** No. 4, 48-51. (Newsletter, North American Thermal Analysis Society).

Rachlin, A. L., Mandarino, J. A., Murowchick, B. L., Ramik, R. A., Dunn, P. J., and Back, M. E. (1986) Mineralogy of hilgardite-4*M* from evaporites in New Brunswick: *Can. Mineral.* **24,** 689-693.

Robinson, G. W. and Chamberlain, S. C. (1984) Famous mineral localities: the Sterling Mine, Antwerp, New York: *Mineralogical Record* **15,** 199-216.

Robinson, N. W. (1968) *The Physical Principles of Ultra-high Vacuum Systems and Equipment:* Chapman and Hall, London, 270 pp.

Rosen, R. T., Hartman, T. G., and Lech, J. (1988) Why not purchase a high resolution magnetic mass spectrometer?: *Mass Spec Source* **11,** no. 3, 13-15. (Newsletter, Scientific Instrument Services).

Smothers, W. J. and Chiang, Y. (1966) *Handbook of Differential Thermal Analysis:* Chemical Publishing Company, New York, 633 pp.

Szekely, T. and Till, F. (1974) New possibilities of resolving complex mechanism of decomposition-reactions by TGA-MS system: in *Thermal Analysis, Proc. 4th Int. Conf. Thermal Analysis, Budapest, 1974, Vol. 3,* I. Buzas, ed., Heyden & Son, London, 917-928.

Van Atta, C. M. (1965) *Vacuum Science and Engineering:* McGraw Hill, New York, 459 pp.

Walker, G. F. and Cole, W. F. (1957) The vermiculite minerals: in *The Differential Thermal Investigation of Clays,* R. C. Mackenzie, ed., Mineral. Soc., London, 191-206.

Webb, T. L. and Heysteck, H. (1957) The carbonate minerals: in *The Differential Thermal Investigation of Clays,* R. C. Mackenzie, ed., Mineral. Soc., London, 329-363.

Weast, R. C., ed. (1980) Constant humidity table: in *CRC Handbook of Chemistry and Physics,* 61st ed., CRC Press, Boca Raton, Florida, E-46.

Wendlandt, W. W. (1986) *Thermal Analysis:* Wiley, New York, 814 pp.

MINERAL INDEX

J. W. STUCKI

Alumina 38
Aluminum oxide 34-35, 38, 52, 85
Amphibole 177
Analcime 143-150
Anhydrite 100, 132
Apophyllite 177
Aragonite 119, 122, 132
Arsenates 179, 184
Arsenites 179
Astrophyllite 177

Barium chloride 142
Bentonite 138
Benzoic acid 34, 36, 38
Biotite 177
Boehmite 133-134
Borates 2, 179
Brucite 138, 140, 180
Brushite 168

Calcite 119, 122, 132, 134, 138, 180, 183
Calcium oxalate 100-102, 112, 143, 179
Chlorite 62-63, 90-91, 112, 117, 172
Chlorite, CCa-1 117-118
Clinoptilolite 100, 124-128, 130-131, 148
Cobalt chloride 142
Cyanotrichite 160-161

Dickite 42, 116, 121
Dolomite 119, 123

Feldspar 132

Goethite 136, 141-143
Gypsum 100-101 103-104, 107-112, 114-116, 132

Halloysite 39-40, 42, 44, 116, 119
Hematite 136, 141
Heulandite 125-127, 131
Hilgardite 178
Hydralsite 78
Hydrotalcite 132

Illite 75

Kaolin 46, 63, 151
Kaolinite 34, 42, 50, 54, 75, 79, 85-86, 88, 91, 101-103, 111, 116, 129, 132-135, 138, 140-141
Kaolinite, API #9 120
Kaolinite, intercalated 24-25, 28-29, 35, 41-45
Kaolinite, KGa-1 24, 27, 35, 42-43, 54, 78, 82-83, 120
Kaolinite, KGa-2 103
Kraisslite 179

Lepidomelane 177

Magnesite 119, 123
Magnetite 102
Maricopaite 174-175
Metakaolinite 75, 79, 83, 85-88, 102
Mica 56, 63, 76, 90, 172, 177
Millerite 179, 180-181
Montmorillonite 63-64, 67, 70, 75-76, 86, 88-91, 134, 142
Montmorillonite dehydroxylate 75, 88, 89
Montmorillonite, SAz-1 124, 129
Montmorillonite, STx-1 124, 127
Montmorillonite, SWy-1 54-59, 64-69, 71-73, 76, 78, 80-81, 84, 89-90, 124-128, 130, 132
Muscovite 56, 76, 90

Oxalate 100-102, 112, 143, 179

Paragonite 90
Phosphate 180, 184
Pyroaurite 132
Pyrolusite 103-104
Pyrophyllite 63, 75-78
Pyrophyllite dehydroxylate 76

Quartz 81, 83-84, 132, 134

Ripidolite 117-118
Romanechite 111, 113

Saponite 63
Scapolites 179
Serpentine 63
Silica 54
Sodium tungstate 97-99
Sulfides 2, 179
Takovite 132-135
Talc 63, 175-177
Titanite 177-178
Titanium oxide 52
Todorokite 111, 113
Tundrite 170

Vermiculite 63, 90, 172-173

Weddelite 178
Wicksite 180

Zeolite 100, 119, 124-128, 130-131, 138, 142-143, 148, 151, 170